Eight essays from Rachel Carson to Agroforestry

Cheryl Lans

Contents

Rachel Carson paved a path of limited advocacy for other scientists............................ 2

Science is politics by other means when it comes to witchcraft........................ 8

Ants, academic politics and integrity.. 23

Co-operatives as an alternative model of social organization 50

On how the media influence Canadian democracy.. 69

Methods in EVM - North America and the Netherlands 75

The moral case for animal welfare for snakes and alligators 106

Eco-alternatives to slash and burn agriculture for T&T 145

Index 170

ISBN 978-0-9880852-4-4

Rachel Carson paved a path of limited advocacy for other scientists

Edmund Russel's book *War and Nature: Fighting Humans and Insects with Chemicals from World War 1 to Silent Spring* (2001) describes how *Silent Spring* energized American environmentalists and led to the ban on DDT ten years later. Russel described how chemical warfare was rejected after World War 1 so the Chemical Warfare Service lobbied the US Congress to use their technology in agriculture as peaceful warfare against pests. Carson called this kind of pest control a self-defeating form of warfare.

If Rachel Carson had survived she would no doubt have been asked by other cancer survivors to continue writing about the connections between organic pollutants and human cancers – the environmental endocrine hypothesis. For example aldrin oxidizes in the environment to form dieldrin which is bioaccumulative; i.e. it does not break down easily in the environment and becomes more concentrated as it moves up the food chain to humans and other wildlife. In 2004 the Stockholm Convention banned twelve Persistent Organic Pollutants (POPs). However the estrogenic effects and potential cancerous properties of the persistent organochlorine pesticides such as dieldrin, endosulfan, and lindane are still studied because exposure to them continues.

Carson would have had to negotiate a role between science and advocacy in order to fight the chemical manufacturers, the people who use their toxic products, the scientists paid to discredit real science as "junk" and the environmental-skeptics. She had already experienced *Reader's Digest* rejection of a series of articles on DDT testing in Maryland. One of Carson's antagonists, the Monsanto Corporation, is more politically-connected than ever. Her book's success was partially due to her sharing it before publication with many scientists and serializing it in *The New Yorker*. Murray Bookchin the advocate, did not use these strategies for his own unheralded book *Our Synthetic Environment* (pseud. Lewis Herber, 1962), about the potential harmfulness of pesticides, food additives, and radiation. In its obituary of April 15, 1964 The *New York Times* wrote that Miss Carson rejected the advocacy label and was only opposed to the indiscriminate use of chemicals.

Carson reminds me of Marie Curie:

She died a famous woman denying
her wounds
denying
her wounds came from the same source as her power - Adrienne Rich, *Power* (1974)

Similarly Rachel Carson testified before Congress while hiding her radiation treatment, after writing that ... "chemicals are the sinister and little-recognized partners of radiation in changing the very nature of the world." However it's Carson's advocacy style that University of Washington Professor James Karr invokes in a (2006) *BioScience* article.

Rachel Carson's book had an impact in the Caribbean. For example the (March 1976) *Surveillance Report* of the Caribbean Epidemiology Centre (CAREC) (affiliated to the Pan American Health Organization (PAHO) and the World Health Organization (WHO)) described how chlorinated hydrocarbon insecticides revolutionized the control of insects of public health and agricultural importance in the 1940s. This first paragraph then continues by stating how these chemicals, including DDT remain in the environment and accumulate in the food chains of various species and how *Silent Spring* the well-known book of Rachel Carson focused public attention on the danger posed by these chemicals.

In an interview conducted in Trinidad in September 1999, Trinidadian naturalist Victor Quesnel told me that *Silent Spring* started the local Conservation movement. Like Carson, Quesnel was a traditionally trained scientist. After his Masters degree in Biochemistry at the University of Toronto, and his Doctorate at Cambridge University (1945 – 1954), he returned to Trinidad and worked at the Colonial Microbiological Research Institute, he then joined the Cocoa Research Unit at the University of the West Indies. Quesnel also became Secretary of the Trinidad Field Naturalists' Club (1953-1959) and restarted meetings which had stopped in 1948. Quesnel was President of the Club from 1986 - 1988.

Quesnel told me that *Silent Spring* influenced the decisions made by Club members. He claimed that only three people in Trinidad and Tobago at that time knew about conservation issues: Andrew Carr (1902-1976), a long-time secretary of the Club, Ludolf Wehekind (1895 – 1964) the President of the Club at that time and himself. Quesnel leaves out one newcomer to Trinidad, the oceanographer William Beebe who had settled into the Simla research station (now the Asa Wright Nature Centre). Beebe published Carson's first essay on eels in his 1944 anthology *The Book of Naturalists.* Next he helped her get the Eugene F. Saxton Memorial Fellowship to complete her first book *The Sea Around Us* (1951) which includes an acknowledgement of his support.

Quesnel: A switch to being a conservation Club would change the nature of the Club. Club members looked on and threw up their hands on some environmental issues. The Conservation attitudes are very recent.... As conservation issues became more pressing the Club got more involved in those issues otherwise the Club members could not enjoy themselves.... The problem was also that there were so many conservation issues; it was hard to focus on any one issue for a long time. That was a defect in a sense, that the Conservation effort was not focused enough. I suggested that we should grade issues and tackle them, that never happened...There was an all day meeting at the Nature Centre to redefine the aims of the Club and the conservation issue was discussed. The conclusion was that the Club's contribution should be education. ..The Scientific Journal, lectures and meetings ..After Silent Spring was published the Club became a member of the Blue River Action Committee working to prevent the transportation of liquid pro-gas by the Pelican barge along the river in the Caroni Swamp. The Club also spoke out against excessive quarrying. It didn't behoove the Club to become a pressure group. Other people are now taking the credit for the work the Club did on leatherback turtles [17 years of tagging and beach patrols during the nesting season at Matura and other beaches until 1980 and outreach to secondary schools].

The Trinidad Field Naturalists' Club began on 10th July 1891under the patronage of the then Governor Sir F.N. Broome and published its first journal in 1892. The Club's motto was *Natura maxime Miranda in minimis* (Nature is greatest in little things). The following men were at this first meeting: Richard Mole (1860 - 1926) who came to Trinidad from Britain as a journalist and naturalist in 1886. He sent specimens to the British Museum and collaborated with the Trinidad-born Friedrich Urich (Jangoons) (1870 - 1937) on a reptile paper that was published in 1891. Jangoons was a bookkeeper, accountant, Inspector of Schools, librarian, Board of Agriculture entomologist from 1909, Department of Agriculture entomologist (1920 – 34), and retired as an Assistant Professor at the Imperial College of Tropical Agriculture.

The Superintendent of Schools Plantagenet Lechmere Guppy (1871 - 1934) was the Trinidad-born son of a British-born civil engineer R.J. Lechmere Guppy (1836 – 1916). He was presenting a paper at the Zoological Society in England and wrote to wish the new Club well. Another founding member Walter Broadway (1863 - 1935) was English but came to Trinidad in 1888 to work at the newly created Royal Botanic Garden. Henry Caracciolo (1859-1934) was the Club's Founding President and a Fellow of the Linnean Society of London. Mr. Alfred Taitt (an Anglican Vicar) was the first Secretary and Treasurer. Thomas Irwin Potter (1864 – 1943) was a Trinidad-born white but his probably-British ancestors had lived in St. Vincent for a century. G W Hewlett was English and left Trinidad for South Africa by 1894.

In these early years the Club members hosted foreign naturalists and explorers including a team led by Dr. Herbert Spencer Dickey (April 1931) looking for the source of the Orinoco River. Jangoons, G.B Rorer, a mycologist and Archer Warner, the Solicitor General, took former US President Theodore Roosevelt to visit the oilbirds (*Steatornis caripensis*) in the Cumaca cave in March 1916. These collaborations resulted in scientific recognition: Henry Caracciolo had a vampire bat named after him *Vampyrops caraccioli* O. Thomas. The guppy fish *Girardinus guppii* was named after R.J. Lechmere Guppy. *Caecorhamdia urichi* (catfish) is one of the species named after Urich, some others are the *Eleutherodactylus, Antipaluria, Liothrips, Calospila, Antiplauria, Hylodes urichi* Boettger, *Sericomyrmex* and *Solenopsis basalis* urichi Forel.

From 1907 - 1925 the Club stopped meeting. Broadway worked at the Grenada Botanical Garden from 1894 – 1914. When the members came together in 1924 to wind the Club up, Andrew Carr stood up and said that History would not forgive them if they allowed the Club to die and he offered himself as Secretary; he remained for 20 years. He was a coloured man in a club started by whites, because it had never been the rule or aim to exclude non-whites. The Club spawned off the Zoological Society (1947), Caribbean Forest Conservation Association (CFCA) (1987) and Environment Tobago (1995). Tobago was added to the name of the Club in 1973. German-born

Catholic Father Leonard Graf joined in 1924 and was President from 1940 to 1945. Father Graf taught Botany and Zoology to the teenaged boys of St. Mary's College and arranged many naturalist trips for the young men of the time (who could not become junior members of the Club). The Club grew to 400 members by 1978. The Field Naturalist quarterly bulletin started in 1973 and they publish a scientific journal called Living World (issues 1891-1896, 1981 to current).

The same pesticides that Rachel Carson wrote about were used in Trinidad to eradicate disease-causing mosquitoes. The scientists working on these efforts were employed at the Insect Vector Control Division, Ministry of Health and Environment, or the Caribbean Epidemiology Centre (CAREC). These scientists were either members of the Club or had to cite the works of the Club's journal. For example T.H.G. Aitken of the Trinidad Regional Virus Laboratory (T.R.V.L.) (later of the Yale Arbovirus Unit) wrote in the Club's publications about sandflies, epiphytes, the Bromeliaceae and arthropods collected by personnel of the Lab (T.R.V.L.). The T.R.V.L was established in Trinidad in 1953 by the Rockefeller Foundation in co-operation with the Trinidad Government. The Virus lab's first Director until 1961 was the epidemiologist Dr. Wilbur Downs. The Rockefeller Foundation was involved in mosquito control in Brazil (1939-41), Mexico (1943-1952) and Italy (1924-35).

DDT was used in Trinidad to control the spread of mosquitoes during the yellow fever mosquito eradication campaign from the 1950s until 1961. *Aedes aegypti* (L.) is the primary vector in the spread of dengue haemorrhagic fever and yellow fever viruses. The viruses came from Africa via the slave trade in the sixteenth century. Yellow fever outbreaks had been documented in the eighteenth and nineteenth centuries but Wilbur Downs published a paper in 1955 stating that there had been only isolated outbreaks in 1907 and 1914. He was writing about cases that were found in 1954 following testing by the T.R. V.L. in collaboration with the Rockefeller Foundation Virus Labs in New York in 1953-54. Carson mentions this 1954 outbreak in *Silent Spring*. The chemicals used next were dieldrin/ aldrin (1948, banned in 1974, 1985) and chlordane (1948 to total ban in 1988). Malathion and diazinon are currently used.

Carson wrote about two children in Florida who used a bag that formerly contained parathion to repair a swing; they died; just like the seventeen Jamaicans who died in 1976 because a batch of European flour became contaminated in a warehouse. Scientists are now searching for rather than testing DDT, finding that DDT-containing mosquito coils (rather than allethrin) were imported into Trinidad from China from1992-1994.

If Rachel Carson had survived she would have repeated that Mankind struggles with the challenge "to prove our maturity and our mastery, not of nature, but of ourselves."... "We still talk in terms of conquest. ...We still haven't become mature enough to think of ourselves as only a tiny part of a vast and incredible universe. .. man is a part of nature, and his war against nature is inevitably a war against himself. "

Science is politics by other means when it comes to witchcraft

(Bruno Latour in Susan Leigh Star 1991)

When I decided that the commodisation of knowledge was a good topic to write about it was immediately assumed that I meant commoditisation as illuminated by Long et al. Apparently these sociologists had a well-known debate in the commodisation area but the focus was mainly on labour in agriculture, rather than knowledge per se. While I was in the process of searching minds rather than the apparently unknowing machines for references on the knowledge-commoditisation topic, the experience of being referred from one male professor to the other and then of being told by the last that the first was the one who knew, (which is why I went to him first), made me decide to focus on the genderisation and commoditisation of knowledge.

For the last decade there has been an explosion of business and management science literature most of which claims quite boldly that knowledge/information is the latest resource, it is a commodity, it needs to be managed. When did knowledge become something that needed to be managed ? In addition the users of knowledge need to be defined so that 'market niches' and 'target groups' can be identified. In this way the 'knowledge generators and disseminators' can 'position' themselves correctly and profitably. The gap between instant information and profit has become very small, in fact many training institutes earn their living in this gap. Knowledge has become less of an action and more of a thing in itself - a commodity.

The most prevalent action associated with knowledge in these times is the constant running to keep up with the latest research, the newest knowledge, so that the knowledge commodity does not become obsolete and unmarketable. Often this implies that the previous knowledge/theories are discarded as 'old news'. Usually it is the groups with the most power who can control which knowledge is 'new' and 'relevant' and thus disseminated.

As a Trinidadian would put it "knowledge as a commodity...but that is stale news". Apparently science has not become a commodity; it always was. Commoditization has developed alongside the growth of the scientific community. It has been part and parcel of science as a professional -and thus autonomous -enterprise. Academics such as doctors and lawyers are shown below trading knowledge and professional expertise for resources, recognition and professional autonomy. What is 'real news' now is the scale of the intervention by government and industrial interests into research/learning institutes. Shiva (199) is concerned not only with the scale of the above mentioned intervention but also by the domination and subjugation of nature and women by patriarchal, reductionist, mechanical science and its arbitrary division of 'knowledge' and 'ignorance'. Fujimura (quoted in Leigh Star 1991) says she is interested in understanding why and how some human perspectives win over others in the construction of technologies and truths, why and how some human actors will go along with the will of other actors, and why and how some human actors resist being enrolled, in this essay I share that interest.

..the scientific method..does not exist in splendid isolation, but is an organic part of a coherent cultural system..Viola Klein (in Spender 1988;699)

This essay focusses on the commoditisation of women's medical knowledge which some feminists say began in the sixteenth century with the great European witch burning. The burning was apparently one of the mechanisms to control and subordinate women, the peasants and the artisans, who in their economic and sexual independence constituted a threat for the emerging bourgeois order, (Mies, 1986:81), some academic women were also burned. They suggest that christianity, capitalism and science are all 'children' of patriarchy and that all existing knowledge and its commercialisation has to be seen in a patriarchal light. What they say goes against all common wisdom, but also makes you question why some wisdom is more 'common' than others, and why some [female] scholars are so little known that feminist researchers can claim that they have been erased from HIStory.

Gage (in Spender 1988;325) puts Church and State as manifestations of patriarchy and weaves together the sexual double standard, the appropriation of women's bodies, the enslavement of women and the practice of witch hunting as a pattern of patriarchy, a framework of systematised oppression of women, all based on male control of knowledge.

The Old English word Wicca, meaning 'wisewoman' is a positive term for maiden, mother and crone. Wicca creates rituals and ceremonies using the symbolism of womanhood. Matilda Gage (Gage, Woman, Church and State 1873 quoted in Spender 198) is credited as the first to describe witches as bearers of an alternative feminine tradition which gave them powers feared by the churches. Originally a midwife was a wise woman or ancient priestess, but historical and material conditions have devalued their role, (Humm ,1990: 236). For example from the tenth century onwards the association between herbs, flowers and the victory of Mary over death was celebrated in the ritual of the Lammas, one of the four great feasts of the Old Religion of the witches. This ritual was discouraged and the date was used for another church ceremony (Daly,1984:129).

...the new knowledge, too, is subject to commoditization and appropriation. But because this knowledge has local origins, somewhere, this process of appropriation has a new twist. It can remain knowledge even while its original knowers can be declared to have lost the competence to know it... (Vitebsky,199 :110)

According to feminists like Mary Daly, Carolyn Merchant and Maria Mies the rise of the medical establishment and the zeal of certain christian churches combined with the strengthening patriarchal culture to create an environment where newly professional men claimed medicine, law, science, philosophy and religion as areas under their exclusive control. Women's natural knowledge in midwifery and natural healing was destroyed by burning the knowledgeable as witches (a precedent to burning books ?) in order for male doctors to become the only experts and to start their journey to god-like status, (the biomedical model of all youth/life-prolonging medicine, the creation of test tube babies), and for the subordination of women and nature to proceed apace. The development of first the forceps, obstetrics-gynecology and a masculine medical science all contributed to the elimination/deskilling of midwives in America, Britain and Europe. Medical practice is now largely based on male morality and ideology...in vitro fertilisation takes precedence over male-located birth control. Women have been removed from an active state of knowing into the passive state of patients and medical knowledge has changed from a locally based heterogenous attribute of midwives/natural healers that developed slowly and cumulatively through daily experience, trial and error, into a commodity localised in institutes where it can be imparted to the 'best minds' for a certain training fee.

...Now the natives need to be re-educated in their own forms of knowledge by those who have become better custodians of it... (Vitebsky,199 :110)

Feminist critics Barbara Ehrenreich and Deirdre English (1979) (quoted in Humm, 1990) among others question the content of medical knowledge and the form in which it is reproduced, they claim it is the most powerful source of sexist ideology in world culture. They show that throughout history women have been continually denigrated by male doctors. Medicine defines sickness as part of women's general condition and treats women's health issues as irrelevant. The sexist medical model is at its most explicit in the institutional control of women's reproduction...what Vandana Shiva calls the takeover of biological reproduction by capital and technology, shifting power from mother to doctor, from women to men. (Shiva 26: 199) Medical reproductive technology and practices rely on machine metaphors, technological intervention and chemical engineering. The medicalization of childbirth has been linked to the mechanization of the female body into a set of fragmented, fetishized and replaceable parts, to be managed by professional experts. A woman's direct organic bond with the foetus is replaced by knowledge mediated by men and machines which claims the monopoly of expertise to educate women to be good mothers.(Shiva 199 ;26). Women until recently could only enter the profession as lowly paid nurses. Where doctors are mainly female as in the former Soviet Union the profession as a whole has less status.

..Knowledge in general, and scientific knowledge in particular, serves two gods: power and transcendence. It aspires alternately to mastery over and union with nature...(Fox Keller 199 ;183)

Fanny Wright writing in the 1800s says the ruling elites 'come selfishly and not generously to the tree of knowledge..they eat but care not to impart of the fruit to others. As quoted by Spender (1988;167) men acquire knowledge for its trimmings, its status, for the edge it gave them over others.. .'all the branches of knowledge, involved in what is called scholastic learning, are wrapped in fogs of pompous pedantry' ...the oppressed should 'break our mental leading strings' and examine closely where our ideas come from and why it is we know what we know. And why it is that women like Harriet Martineau who wrote books on economics for the lay man without the 'trimmings' have practically 'disappeared' from HIStory. Spender (1988;174) explains it thus..'little respect is given to those who do not play the 'mystifying' game and who instead insist on the knowledgeability of all human beings and who attempt to make knowledge accessible to them.' Dorothy Smith (Spender 1988;9) says 'all human beings are constantly engaged in the process of describing and explaining, and ordering the world, but only a few have been, or are, in a position to have their version treated as serious, and accepted...these the 'circle of men' -who are the philosophers, politicians, poets and policy-makers -who have for centuries been writing and talking to each other about issues which are of significance to them.

One of this 'circle' Aristotle believed that women's consignment to inferior status was neither a punishment or injustice. It was simply a matter of observable fact that, happily, supported the existing social order. His doctrine was carried over with some modifications into Christian doctrine. The notion of the woman's body as nurturant host to the paternally provided embryo persists to this day, despite the 16th-century discovery of the ovum and the modern understanding of genetic dimorphism. But rhetoric and "common" sense agree that men have children "by" (i.e., through the instrumentality of) women, just as race horses are "sired by" male stallions "out of" mares (Hein, 296).

According to Spender (1988;7) 'the suggestion that men have omitted fundamental questions about the nature of existence in their years and years of checking with each other, is a preposterous one in a patriarchal world. It casts aspersions on the accumulated wisdom of men, and their status as authority figures. It is therefore a suggestion which will be met with considerable resistance, and many are the mechanisms the ruling group has made available to discredit and devalue any member of the 'ruled' who dares to put it forward. Spender continues 'Patriarchy requires that any conceptualization of the world in which men and their power are a central problem should become invisible and unreal. How could patriarchy afford to accept that men were a serious problem, to gather together what women have said on the subject and make it the substance of the educational curriculum, to treat women's version of men with respect, and to discuss at length and with impartiality the origins of male power and the means by which it is maintained ? This is a nonsense in a patriarchal society. It is akin to asking christianity in the past to put favorably the case of heretics, or the scientific community in the present to devote an equal amount of time to address with equal good will such areas as alchemy, faith healing, astrology (or even the case against nuclear power) which it has outlawed, and against which it sets itself up as authoritative.' (Spender 1988;9)

...When, however, a knowledge is exported beyond the limits of its area, it can survive and be sustained only by some claim to universality. Often in 'world religions'...this is done militantly. Then it must annihilate, degrade or subsume other ways of knowing, seeing these as rivals in a colosseum of mutually negating knowledges... For the triumphant party, one's knowledge is no longer local; for the defeated, it is no longer knowledge. Within these, there are many degrees of resistance, coercion and collusion. The destruction of local knowledge creates a class world whose overlords require a proletariat:since there are the landowners of knowledge, there must also be the landless labourers and the unproductive beggars. (Vitebsky 107:199)

According to Mies (Mies, 1986: 70) the male minds behind the witch hunt denounced female nature as sinful and sexually uncontrollable. She claims that sexual autonomy is closely connected with economic autonomy, when the church united with the newly professional doctors to drive out and denounce women healers and midwives as witches Mies claims this as an example of the onslaught on female productive activity which also served to help establish male autonomy over all spheres.

Mary Daly (1984) describes the torture and killing of the witches as a conspiracy theory with men as the evil force and woman as basically good. She describes the attack on women's agency as taking place on all fronts, even symbolic. Daly claims that the symbol of Mary Mother of God contained a lingering metaphoric link reminding women of the potency in women and nature and women's elemental power. The image of Mary had metaphoric power because she was a remnant of what was alive and thus evoked memories of what had been alive, that is, the Goddess symbol in women's consciousness, (Daly,1984: 92-110). The Catholic church needed to maintain Mary which was still an important symbol to the masses but as Daly says because 'women sensed in the presence of Mary the elemental powers in themselves and their sisters ..."Mary malfunctioned as an archetype; the situation was intolerable to the mind-molders", so both the image of Mary and the healers/witches themselves were destroyed by the Protestants.

The Catholics kept the Mary image but "drained it of its residual vibrancy and remolded it into an archetypal shape". During the sixteenth century Reformation Protestant churches replaced the symbol of Mary with a Jesus that "incorporated both masculine and feminine roles, being lord, saviour, and sacrificial victim". Protestant reformers also closed female monasteries, forcing nuns to leave their autonomous condition and be assimilated into marriage. Gage (in Spender 1988;325) says women were excluded from the church from the fourth to ninth centuries, first ordination, then deaconship, then serving at the alter.

Matilda Joslyn Gage predated and influenced Daly. (Spender 198 :327). Christianity in her view did not lead to man's treatment of women but grew out of it, reinforced it and extended it, (ibid., p.319). She claims the doctrine of original sin was introduced by Augustine and the Church found it 'attractive, self-enhancing and profitable'. Once the 'contaminating' nature of women was established, they became a source of impurity and contamination, so, men who had no contact with women were 'purer', and the clergy became celibate. Women were then excluded and avoided, and the "'purer' sex, closer to god," and removed from original sin made woman accountable to him the same as he was accountable to God. Gage says the Church also provided the logic and rationale for mens' ownership of women's labour. The Church declared woman to be inferior and men were then free to steal women's creative and physical and intellectual energy. Using circular reasoning the Church then stood vindicated for it became obvious after the 'theft' that women were indeed without resources and therefore inferior. Since the knowledge was not allowed to belong to women it was not theirs to pass on and the knowledge was appropriated, ignored or suppressed and its creators 'disappeared'.

...when we look for women's view of the world, from the position
of 'the weak', or the oppressed, from the underside, we encounter little but silence...(Spender 1988;12)

Mies proposes that the attack of church and state against the witches was aimed not only at the subordination of female sexuality as such but against their practices as abortionists and midwives. Women resisted by joining heterodox sects in which they played a prominent role, or which in their ideology propagated freedom and equality for women and a condemnation of sexual repression, property and monogamy.

Thus the "Brethren of the Free Spirit", a sect which existed over several hundred years, established communal living, abolished marriage, and rejected the authority of the church. Many women, some of them extraordinary scholars belonged to this sect. Several of them were burnt as heretics. The institutional and ideological props necessary for the maintenance of women's future self-repression were provided by the church, the state, and through the family (Mies,1986:81).

..To speak for others is to first silence those in whose name we speak. (Callon 1986:216 in Susan Leigh Star 1991)

Hein says ...the deliberate taking of life is a transcendent act. To impose death (As distinct from suffering it) is to declare oneself a self. Thus men have glorified death, awarded themselves authority over it, and mystified it where their power to control it was lacking... (Hein 199; 299).

Gage according to Spender (1988) speaks of 'the most stupendous system of organized robbery known has been that of the church towards woman, a robbery that has not only taken her self-respect but all rights of person...she has no right to live for herself alone..her position must always be secondary even to her own children, her right to life has been admitted only insofar as its reacting effect upon another could be predicated."

Mies (199: 146) credits Merchant (1983) with the theory that "only after the witches had been killed as 'bad women' could a new image of the 'good woman' emerge in the eighteenth and nineteenth centuries. This was the image of the vapid, sentimental, weak, oppressed woman, the woman as counter image and dependent on the modern, rational breadwinner and the state (what Daly calls the living death of patriarchal marriage). This new bourgeois ideal of womanhood was necessary for the new sexual and social division of labour, the division between production and reproduction, production and consumption, work and life, needed for capitalism". Mies describes how Steinbrigge depicted the 'feminine' image construction as the brainchild of Enlightenment philosophers, Diderot and Rousseau, a last bastion of humanness in the emerging society of male reason.

..part of the process whereby knowledge can become totalitarian is to draw its interlocutors away from this active mode and to convert them into objects: to diminish their agency and consequently their awareness..., (Vitebsky,199 :112).

At least one scholar (Lipshitz 41:1978) says to reclaim women witches or hysterics as examples of lost female power or lost healers, and to search for proof of their existence, as has some feminist literature, is to take them too seriously. She says that witchcraft and possession by spirits are forms of illness,   these women are acting out their frustrations with their gendered roles by claiming to be possessed. But, Mies also presents evidence that the Witchcraze in Europe was not an offshoot of the dark and superstitious Middle Ages but was fostered and fuelled by the rising professional power block...the priests and ministers, and the emerging legal and medical professions all had an active role and vested interest in the killing of the witches.

Spender quotes Gage claiming no coincidence in the fact that 'at the time witchcraft became the great ogre against which the church expended all its terrific powers, women doctors employed anaesthetics to mitigate the pains and perils of motherhood'. 'The church having forbidden its offices and all external methods of knowledge to woman, was indignant at her having through her own wisdom penetrated into some of the most deeply subtle secrets of nature.'

Not only were women able to own some knowledge, says Spender, not possessed by men, not only were they able to act in a way in which their existence was unaccountable to men, but they were using their power to alleviate some of the pain which the Church insisted women must suffer as a form of penance....an intolerable subversion of the patriarchal world'(Spender198 :328).   Merchant (1983) calls the holocaust of women part and parcel of the beginning of the New Modern Age. Daly (1984) puts the number of women killed at nine million, in Protestant and Catholic countries. The Church 'had to erase women with the power to heal, not only by killing them, but by denying that they healed of their own power. The only way to sustain the belief that women had no resources of their own was to insist that this power originated from the male enemy of god/man..the devil.

Jean Bodin, (Merchant 1983:138) the French theoretician of the new mercantilist doctrine...decided that for the development of new wealth after the medieval crisis, the modern state had to be invested with absolute sovereignty. He was the founder of the quantitative theory of money, of the modern concept of sovereignty and of mercantilist populationism, the state had the duty to provide for enough workers for the new economy. To do this a strong police force was needed to fight against witches and midwives who he felt were responsible for abortions, infertility of couples and sexual intercourse without conception. He was a consultant of the French government in the persecution of the witches and advocated torture and the pyre to eradicate the witches. His tract on witchcraft was one of the most brutal and sadistic of all pamphlets written against witches at that time.

......the relationship of people to this knowledge, too, becomes a form of consumerism. This consumerism extends not just to knowledge as a commodity, but also to wisdom...(Vitbesky,199 :110)

According to Gage quoted in (Spender 198 :331) 'the three most distinguishing features of the history of witchcraft were its use for the enrichment of the church;for the advancement of political schemes;for the gratification of private malice'. Spender continues that 'it created a whole industry which provided jobs for men: the clerics who interpreted the scriptures, the learned judges, the accusers, torturers, witch-prickers.

..trials for witchcraft filled the coffers of the church, as whenever conviction took place the property of the witch and her family was confiscated to that body. The clergy fattened upon the torture and burning of women (Gage in Spender 198 :329)

Mies also shows a direct connection between the witch pogroms and the emergence of the professionalization of law (Mies, 1986:84). Before that period, the German law followed old Germanic custom; it was people's law or customary law, but not a discipline to be studied. But now Roman law was introduced, most of the universities established a law faculty or had only that faculty. The witch trials provided employment and money for these new lawyers and the legal profession, especially since according to Canon Law the property of the witch was confiscated and at least 50% was appropriated by the government. This 'political economy' of the witch hunt also raised funds for the warring European princes to finance the Thirty-Year War from 1618-1648. The lawyers of Cologne argued that the witches received a lot of money from the devil and that it was perfectly in order that this devil's money be confiscated by the authorities in order to eradicate the witches (Mies, 1986:86).

Merchant has demonstrated a link between the torture of the witches and the rise of the new empirical scientific method; the destruction of the integrity of both the female body and the body of Nature (Merchant,1983:164-177). Both were to become mere sources of raw material for the rising capitalist mode of production. According to Mies, (Mies, 1986:69) the witch hunt in Europe was the first violent subordination of women under men in the process of capital accumulation. Women and colonial peoples were defined as property, as nature, not as free subjects, who could enter a contract. However they had their own will which had to be subordinated by force under the will of the 'free' subjects of civilised society, the men, as well as under the law of capital accumulation. Women did not voluntarily hand over their productivity, their sexuality and their generative capacities to their husbands and to what Mies calls the BIG MEN (Church, State) Both had to be subordinated by force and direct violence. This violence constituted the infrastructure upon which so-called capitalist production relations could be established, namely the contractual market or relinquish their knowledge to an outside expert, they have to be forced to produce things they do not consume themselves. Male guilds and the rising urban bourgeoisie pushed craftwomen out of the sphere of production. Man cannot produce living human labour therefore the generative and productive forces of the European women had to be brought under control.

As stated above the interrogation of witches also provided the model for the development of the new scientific method of extracting secrets from Mother Nature. Carolyn Merchant (1983) has shown that Francis Bacon, the 'father' of modern science, the founder of the inductive method, used the same methods, the same ideology to examine nature which the witch-persecutioners used to extract the secrets from the witches, namely, torture, destruction, violence. He deliberately used the imagery of the witch hunt to describe the new scientific method: he treated nature as female to be tortured through mechanical inventions (Merchant,1983:168) as the witches were tortured by new machines. ...so nature exhibits herself more clearly under the trials and vexations of art (mechanical devices) than when left to herself. he stated that the method by which nature's secrets might be discovered consisted in investigating the secrets of witchcraft by inquisition: for you have but to follow and as it were hound out nature in her wanderings, and you will be able when you like to lead and drive her afterward to the same place again..." (quoted by Merchant, 1983:168).

Bacon strongly advocated the breaking of all taboos which, in medieval society, forbade the digging of holes into Mother Earth or violating her: Neither ought a man to make scruple of entering and penetrating into these holes and corners, when the inquisition of truth is his whole object (Merchant, 1983:168). According to Bacon, nature must be 'bound into service", 'made a 'slave', put into constraint', and to be dissected; much as a woman's womb and symbolically yielded to the forceps, so nature's womb harboured secrets that through technology could be wrested from her grasp for use in the improvement of the human condition.

...a set of uncertainties are translated into certainties: old identities discarded, and the focus of the world narrowed into a set of facts...(Leigh Star 1991)

Bacon's scientific method, which is still the foundation of modern science, unified knowledge with material power. Merchant concludes that the interrogation of witches as symbol for the interrogation , and that women be forcibly made to breed more workers. Similarly, nature had to be transformed into a vast reservoir of material resources to be exploited and turned into profit by this class. Hence the state, the church, the new capitalist class and the modern scientists collaborated in the violent subjugation of nature and women.

...to translate is to displace...but to translate is also to express in one's own language what others says and want, why they act in the way they do and how they associate with each other; it is to establish oneself as a spokesman. At the end of the process, if it is successful, only voices speaking in unison will be heard. (Callon 1986:223 in Leigh Star 1991)

The social upheaval of the times was partly due to the decaying old order in its confrontation with new capitalist forces, but was also a reaction of the new male-dominated classes against the rebellion of women against their expropriators. When a woman denied being a witch and having anything to do with all the accusations, she was tortured and finally burnt at the state. The witch trial, however, followed a meticulously thought-out legal procedure, (Mies, 1986:84). In protestant countries one finds special secular witch commissions and witch commissars. The priests were in constant rapport with the courts and influenced the judges.

...During torture (and in similar ways during war) the world is created and uncreated. The torturer shrinks the world of the tortured, by taking the uncertainty of experienced pain and focussing it on material objects and on the verbal interchange between them. Old identities are erased made immaterial...In torture, it is in part the obsessive display of agency that permits one person's body to be translated into another person's voice, that allows real human pain to be converted into a regime's fiction of power (Scarry 1988:18 in Leigh Star 1991)

In 1631 Friedrich von Spee dared to write an anonymous essay against the tortures and the witch-hunt, (Mies, 1986:82). He exposed the sadistic character of the tortures and also the use the authorities, the church and the secular authorities made of the witch hysteria to find a scapegoat for all problems and disturbances and the unrest of the poor people, and to divert the wrath of the people from them against some poor women. The persecution and burning of the midwives was directly connected with the emergence of modern society: the professionalisation of medicine, the rise of medicine as a 'natural science', the rise of science and of the modern economy. The torture chambers of the witch-hunters were the laboratories where the texture, the anatomy, the resistance of the human body -mainly the female body - was studied.

..we create metaphors...bridges between different worlds. Power is about whose metaphor brings worlds together, and holds them there...metaphors may heal or create, erase or violate, impose a voice, or embody more than one voice. (Leigh Star, 1991)

According to Vandana Shiva 199 "the construction of 'intellectual property' is linked to multiple levels of dispossession. At the first level, the creation of the disembodied knowing mind is linked to the destruction of knowledge as a commons. In contrast to the organic metaphors, in which concepts of order and power were based on interdependence and reciprocity, the metaphor of nature as a machine was based on the assumption of divisibility and manipulability."

"Uniformity permits knowledge of parts of a system to stand for knowledge of the whole. Divisibility permits context-free abstraction of knowledge, and creates criteria of validity based on alienation and non-participation, which is then projected as 'objectivity'. 'Experts' and 'specialists' are thus projected as the only legitimate seekers after and producers of knowledge.

While writing this essay I have learnt that "those who control our accounts of the past exert a great deal of control over our understandings of the present and our visions of the future. And that certain practices of science are cultural artifacts that arise from the gendered symbolism and social structure of science and the masculine identities and behaviors of individual professional men and scientists " (Arnault, 1989: 502)

Bibliography

Daly, M. (1984) Pure Lust: Elemental Feminist Philosophy, Beacon Press: Boston, Mass.

Ehrenreich, B. and English, D. (1979), For Her Own Good:150 years of the Experts' Advice to Women, Anchor Press: Garden City, NY

Humm, Maggie, 1990 The Dictionary of Feminist Theory, Ohio State University Press

Fox Keller, Evelyn 1996, "Feminism and Science", in *Feminism and Science*, E F Keller and H Longino (eds.) Oxford: Oxford University Press: 28–40.

Hein, Hilde, "Liberating Philosophy: An End to the Dichotomy of Spirit and Matter," in *Women, Knowledge, and Reality: Explorations in Feminist Philosophy* Ann Garry and Marilyn Pearsall, eds. (Boston: Unwin Hyman, 1989)
Humm, M. (1990) The Dictionary of Feminist Theory. Columbus, Ohio State University Press.

Star, S. L. (1991) `*Power, Technologies and the Phenomenology of Conventions: On Being Allergic to Onions*', in J. Law (ed.) A Sociology of monsters : essays on power, technology, and domination (1991), pp. 26-56

Susan Lipshitz, "*The Witch and her Devils*",Tearing the Veil, Routledge & Kegan Paul Limited, London,. 1978, p. 42-43

Carolyn Merchant, The Death of Nature: Women, Ecology, and the Scientific Revolution, San Francisco: Harper & Row, 1980

Mies, M. (1986). *Patriarchy and accumulation on a world scale: Women in the international division of labour*. London: Zed Books.

Mies, M., & Shiva, V. (1993). *Ecofeminism*. Halifax, N.S: Fernwood Publications.

Spender, D. (1982). *Women of ideas and what men have done to them: From Aphra Behn to Adrienne Rich*. London: Routledge & Kegan Paul.

Vitebsky, P. (1993), '*Is Death the same Everywhere? Contexts of Knowing and Doubting*', in M. Hobart (ed.), An Anthropological Critique of Development: The Growth of Ignorance, Routledge, London

Henri Rousseau

Ants, academic politics and integrity

Abstract

People typically cannot choose a moral or ethical stance outside of the normative constraints of society at large but have to stay within the norms of the community. Unless of course they are using praxis research to guide society to a path that they consider to be more sustainable, ethical and moral which can be proved to benefit the whole. Integrity in this scenario is individual behavior that improves the lives of everyone in society (sound citizenship).

Keywords: integrity; praxis; Millennium Development Goals; sustainable technology; regimes of truth

I

Benhabib (1987) distinguishes between two types of human strivings: 'the politics of fulfillment which continues the universalist promise of bourgeois revolutions in demanding justice, equality, civil rights, democracy and publicity and second the politics of transfiguration (utopia) which continues the tradition of early socialist, communitarian and anarchist movements seeking qualitatively new relations including happiness and solidarity among self, nature and concrete others'.

Robert Papazian was executed in 1982 with other political prisoners after taking part in the 1979 Iranian Revolution. He lived a life of integrity and died for his beliefs in working class liberation.[1] His parting words to fellow-prisoners were: 'It's not the number of years that counts but the effect of one's life and death on others... Life in a broader sense continues...' Integrity is the daily task of acting in accordance with our values (trust, honesty, responsibility,

---

[1] [online] Available at:http://www.iranrights.org/english/memorial-case--5136.php. [Accessed on August 21 2012].

ethics), attitudes (fortitude) and beliefs (suitably modified by experience) (Bertram, 1997). This includes doing what is right even when it is difficult, or where there is a risk of loss or in other words 'standing for something' (Van Willigenburg, 2000). A person cannot choose a moral or ethical stance outside of the normative constraints of society at large. Instead one has to exercise one's best judgment within the norms of the community (Cox et al., 2012).

The injunction to avoid causing suffering, emotional deprivation, and many other forms of harm is a principle of morality. Harm is bad in itself, not just bad for members of a certain species or type of individual; and not because the individual harmed is or is not a moral person (Beauchamp, 1999). Integrity thus provides part of the foundation for what Schmidtz (1997) calls a 'morality of interpersonal constraint', which forms the basis of institutional arrangements to regulate social interactions between individuals pursuing their personal goals. Integrity is further explained as having an 'integrated self' such that there is consistence between one's commitments and between one's self-embraced and wholeheartedly accepted principles and actions (Van Willigenburg, 2000).

Gray (2002) argues that humans are not different from animals in many respects. Wolfe (1998) points to many studies and language experiments showing that animals have attributes thought to be human- tool making and use, social behavior, altruism, non-verbal language, dislike of boredom, environmental curiosity, parental love, fear of attack, deep friendships, a horror of dismemberment, a repertoire of emotions, exploitive violence, self-awareness, engagement in both deceptive and altruistic behavior, and other qualities. Elephants have a distinct matriarchal culture. As humans encroach on elephant territory and kill them, elephants in turn have been exhibiting cultural breakdown (Bradshaw et al., 2005; Siebert, 2006).

An experiment with two capuchin monkeys, Sammy and Bias, showed that they understood and corrected inequity. This ability is said to underlie human morality (Brosnan and de Waal, 2003). Grey is incorrect to dismiss human morality as a construct since primates understand inequity. Perhaps insects have their own moralities built-in (Bekoff and Pierce, 2009). Social insects live organized, purposeful lives like some human hierarchical societies. Ant colonies usually contain workers, soldiers, a queen and mating partners (Moffett and Tobin, 1991). Bacon (1620) describes ant as empiricists, who 'merely collect things and use them.' In fact ants clean the forest floor, tend fungi gardens and build empires. An Aesop fable warns philosophers not to set human lives above those of ants[2]. Schopenhauer[3] ignored this warning and compared the conflict and strife of human will to a story in which the dissected halves of an Australian bulldog ant attacked each other. Obviously the ant-halves had the will to live but not the wits to recognize itself in the other. Second, in some of the cases described the other members of the ant society carried the ant-parts away. Third while not meaning to quibble too much over a practical, eye-witness account, it was in fact a boy's story[4].

Grey (2002) claims that technological progress is not always linked to moral concerns. I will elaborate on this later in the essay in the discussion of the UN Millennium Development Goals. Grey's book was published before the recent work on the differences in the way that conservative and progressive brains perceive ethics, morality and integrity. This work reveals why human progress seems to take two steps back for each step forward. Lakoff and Wehling (2012) postulate that progressive brains value a large public sector that takes responsibility for the welfare of all; which necessitates collecting taxes to provide for infrastructure, health, justice and environmental care among other concerns. Conservatives on the other hand perceive themselves as self-made men and want minimal taxation for a minimum government that would not have the capacity to stand in the way of their own self-interests. They believe in the authority of a 'strict father' to keep evil human natures in check and in economic austerity for the undeserving poor; while the rich have been favored by God and should not have their wealth taken away.

Lakoff and Wehling (2012) call both of these moral views with consequences for how public policy is shaped. But others have shown that conservatives are benefitting from government social services while wanting to deny these to so-called undeserving others (DeBrabander, 2012). Many of the services that have been recently privatized are run by those donors who are loyal to conservative causes (Farazmand, 1999). George W. Bush wanted to transfer welfare to faith-based organizations and to women, increasing their burdens or forcing them to leave paid

---

[2] [online] Available at: http://mythfolklore.net/aesopica/oxford/index.htm [Accessed on August 21 2012].

[3] Arthur Schopenhauer 1958: *The World as Will and Representation*, Vols. I and II, translated by E. F. J. Payne, New York: Dover Publications (1969).

[4] Howitt, William. 1854, *A boy's adventures in the wilds of Australia, or, Herbert's note-book*. Arthur Hall, Virtue, London.

employment; he used the rhetoric of social justice and moral responsibility (Eubanks, 2009; Colarelli, 2002). A few conservatives are removing the 'liberal translations' from the Bible.[5] While Grey (2002) writes that human nature does not change, these examples show that when conservatives are in power and can set policy they try to reverse social gains (with the help of some moderates and centrists).

II

Research has discovered that conservatives are more fearful and defensive and have an increased volume of the right amygdale part of the brain (fight or flight response, individual survival) while liberals have more grey matter volume in the anterior cingulate cortex (helps manage complexity, group survival) (Kanai et al., 2011; Dodd et al., 2012). A Yale study provides further evidence for the effect of this dichotomy of brains on public policy. The Cultural Cognition Project conducted at the Yale Law School by Professor Kahan and others, with a national sample of 1500 U.S. adults, found little apathy or ignorance about climate change. As members of the public learned more, the individuals belonging to opposing cultural groups (egalitarian/liberal versus individualistic/conservative) disagreed on the risks attributed to impending climate change[6]

> In effect, ordinary members of the public credit or dismiss scientific information on disputed issues based on whether the information strengthens or weakens their ties to others who share their values.. individuals with higher science comprehension are even better at fitting the evidence to their group commitments. (Kahan et al., 2012)

In his *Theses on Feuerbach* Marx outlines how the knowledge claims of humans are based on a practical engagement with nature and states further that ideological conceptions of nature are a product of social life (Lysaker, 2011). When Grey (2002; 2004) claims that scientific progress and religion are delusions because human nature does not change, this is false. Even the 'fundamentalist minority shaping the political agenda' (Grey, 2004) wants change or a reversion to the *status quo ante* that benefits their cultural group. In 2003 Congressman Henry Waxman released a report stating that the Bush Administration manipulated the scientific process and

---

[5] [online] Available at: http://conservapedia.com/Conservative_Bible_Project. [Accessed on August 21 2012].

[6] ; [online] Available at: http://current.com/132qskc; http://www.law.yale.edu/news/15546.htm [Accessed on August 21 2012].

distorted or suppressed scientific findings for the benefit of its financial backers (Waxman Report 2003; see also the Union of Concerned Scientists-Scientific Integrity in Policy Making July 2004).

Colarelli (2002) has a different view of the conservative viewpoint. Conservatives claim that a successful complex society has evolved and that social change and reform may bring about unanticipated and negative changes, therefore social change, if implemented at all, should be done incrementally. Liberals use this exact argument to protest against environmental changes which conservatives favor and consider many human social traditions to be oppressive and in need of change to advance human welfare. Cohen (2011) makes a distinction between small c conservatives who value what already exists more than potential value and large C or movement conservatives who conserve injustice and sacrifice existing things that small c conservatives cherish in order to extend their own wealth and power.

Bertram (2007; 2012) has his own dualistic division. Those who think it is fine to exploit the letter of the law, or to stay just within the rules in order to gain strategic advantage (evading taxes for example), and those who don't agree with this technically legal behavior7. Bertram suggests punishment for the technically innocent but morally culpable so that with each legal defeat of the tricksters, the general population gradually learns to respect the spirit of the law (see also Doring, 1895). Legislators, regulators and sports administrators devise systems of rules over time or within committees that may be incomplete or biased, therefore ex post facto law can be justified. This follows Doring's idea of a progressive accumulation of inherited habits that would in future 'inaugurate a steady, progressive ennoblement of human impulses.' Robert Papazian may have experienced this situation of dueling moral frameworks; his commitment to working class struggle versus the ideas of the traditional religious forces that took power, with the dependence on oil as a critical factor in the conflict.

III

Grey (2004) writes that 'conflict in the Middle East has an extremely complex history, but

---

7 [online] Available at:http://crookedtimber.org/2012/08/02/badminton-taxes-regulation-and-the-peoples-justice/ [Accessed on August 21 2012].

anyone who tells you that western intervention in the region has nothing to do with oil is a fool or a liar.' Two dozen multimillionaires and energy companies gave $54 million in donations to Republican 'super PACs' in 2012 (Ericson et al., 2012). In turn Republicans in Congress shape foreign and domestic policies to benefit these donors. This lack of integrity (correct governance) is bolstered by corporate acts (poor fuel economy standards) and individual acts (poor choice of cars, decisions not to use public transport, decisions to build sprawling urban areas). An ethics for a finite world was published ten years ago in an academic forum:

> Moral codes, no matter how logical and well reasoned, and human rights, no matter how compassionate, must make sense within the limitations of the ecosystem; we cannot disregard the factual consequences of our ethics. If acting morally compromises the ecosystem, then moral behavior must be rethought. (Elliott and Lamm 2002)

Ross (2008) found that oil-dominated economies in the Middle East, Azerbaijan, Russia, Chile, Botswana and Nigeria had repressive gender policies because of the oil wealth, which boosted male-dominated construction (golf courses) and services but suppressed alternative manufacturing which served as career entry points for women (like sewing). Dictatorial regimes in Africa have no development agenda (Adejumobi, 2006). Blaydes and Linzer (2007) found that the resulting lack of economic opportunity in these countries led women to embrace fundamentalist belief systems in order to make themselves more marriageable (their main economic support).

The billionaire Koch brothers use their petrodollars to fund their own advocacy groups to try to achieve their vision of social reversal (Americans for Prosperity, Heartland Institute, Cato Institute, Institute for Humane Studies, American Enterprise Institute, Independent Women's Forum and George Mason University). The Koch-funded Mercatus Center at George Mason University writes deregulation policies for politicians, often aimed at the Environmental Protection Agency, because Koch Industries is one of the top ten US air polluters (Mayer, 2010). The Independent Women's Forum opposes the presentation of global warming as a scientific fact in American public schools (Mayer, 2010). Willie Wei-Hock Soon, an astrophysicist at Harvard testified before Congress in 2003 that global warming was caused by solar variation.

He said that he had 'not knowingly been hired by, nor employed by, nor received grants from any organization that had taken advocacy positions with respect to the Kyoto organization or the UN Framework Convention on Climate Change'. He had in fact received over a million dollars from the coal and petroleum industries (Koch Foundation, Southern Co., Exxon Mobil, American Petroleum Institute, Mobil Foundation, and the Texaco Foundation) (Vidal, 2011).

Habermas' call for neutralized power differences to allow for the creation of consensus through discourse seems naïve in view of these sums of money (Flyvbjerg, 1998). The viewpoint of Foucault seems more correct –

> truth isn't outside power, or lacking in power...it is a matter of knowing what effects of power circulate among scientific statements and why at certain moments a regime undergoes change.... and the process by which a society creates regimes of truth according to its own beliefs and values. (Foucault, 1980)

IV

Francis Bacon also wrote about power and knowledge

> Therefore those two goals of man, *knowledge* and *power*, a pair of twins, are really come to the same thing, and works are chiefly frustrated by ignorance of causes. The whole secret is never to let the mind's eyes stray from things themselves, and to take in images exactly as they are. (Bacon ([1620] 1994)

Bacon expected that controlled experiments with ideal equipment would lead to bias-free scientific observations. Instead some scientists like Emil Abderhalden found many collaborators who found what he wanted them to find, while some scientists who could not replicate what wasn't there had career advancement problems (Deichmann and, Müller-Hill, 1998). Feminists concluded that Bacon's mechanistic view of Nature, whose secrets needed to be harassed-out, had led society in the wrong direction and contributed to gender inequality (Merchant, 2006). Pollack (1992) found that Bacon's path led scientists to create a 'myth that their instruments and procedures somehow free them from the boundaries of their minds and bodies'.

Plato was the first to draw a distinction between the divine soul and the animal body (Binde, 2001). The stoics considered that only man had reason which made him master of soul-less animals. This distinction between animal body, divine soul and mind, and nature was incorporated into early Christianity by St. Augustine and St. Gregory of Nyssa and then further developed by Thomas Aquinas. Later on St. Francis of Assisi and St. Bonaventura introduced a more benign view of pristine nature as a means to approach God. The 'human exemptionalism paradigm' that was further developed in Enlightenment thinking, allows some scientists to claim that they can remake or replace nature with technology and thus transcend nature (Goldman and Schurman 2000; Grey 2002; 2004). However many believers in progress are not seeking salvation from themselves as Grey (2004) states, but salvation from others. For example cellphones in some African countries provide communication and business opportunities that could not be obtained from traditional nation-based phone technology or from foreign aid. 'Gertrude Kitongo uses hers as a radio, library, mini cinema, instant messenger and bank teller. She even makes calls on it'[8]. This technology reduces the economic and environmental burden on African countries that no longer have to provide physical forms of this infrastructure and these institutions in many war-torn locations (Angola, Mozambique, Sudan, Somalia, Rwanda, Burundi, Chad, Sierra Leone, Congo and Liberia). This is a case of technological progress that could lead to changes in the way international development is envisaged.

It is true that there is pervasive video surveillance to stop crime and 'terrorism' as Grey writes (2002; 2004); but in certain instances the public has been able to use this same video evidence to provide proof of police brutality. See the provided links for proof[9].

------

[8] [online] Available at:http://mg.co.za/article/2011-11-09-africa-is-fastest-growing-cellphone-market [Accessed on August 21 2012].

[9] [online] Available at:http://dissenter.firedoglake.com/2012/07/22/officers-offered-to-buy-cell-phone-footage-of-anaheim-police-brutality/ [Accessed on August 21 2012]; [online] Available at: cosmodaddy.wordpress.com/.../manchester-police-brutality-on-cctv/ [Accessed on August 21 2012]; [online] Available at:
http://www.telegraph.co.uk/news/newsvideo/7982843/Shocking-police-brutality-caught-on-CCTV.html [Accessed on August 21 2012]; [online] Available at:
http://www.stamfordadvocate.com/local/article/Stamford-drug-bust-raises-police-brutality-claim-3737528.php [Accessed on August 21 2012].; [online] Available at:
www.news.com.au/...old/...police.../story-e6frfkvr-1226271283171 [Accessed on August 21

Twitter may have been invented by and for business but it has been used in very subversive ways, for example during the Arab Spring uprisings, Small cliques of people with similar ethical and moral standpoints have shared and publicized ideas, events and statements in seconds, thwarting government controls (Lawrence, 2010; Huang, 2011).

V

Grey (2002; 2004) claims that Christianity was created to give human lives meaning. Pre-Christian religions also provided meaning. Some aspects of the story of Jesus and Mary are similar to pre-Christian Egyptian stories of Isis and Horus. Many pre-Christian traditions were incorporated into Christianity (Griffiths, 1980). The Roman emperor Constantine adopted Christianity in order to unite the Roman Empire into a true unit rather than continue to govern a collection of disparate groups (Lunn-Rockliffe, 2011). Christianity provides the foundation for law and social norms in many parts of the world (Bracton, 1968). So it is incorrect for Grey (2004) to claim that there has been no human progress because the Christian ideas of integrity and morality that underpin legal systems in many countries give the laws a commonly-agreed-to framework that makes everyone subject to the law and prevents the worst abuses and arbitrary exceptions for powerful people that would otherwise take place (Bracton. 1968). This is responsible for the complaints that arise when legal decisions seem to stray from what is considered common law justice.

Early Christians were killed for their beliefs and they died thinking that Jesus had also shared their sacrifice on the cross (1 Corinthians 15: 3-4). These early sacrifices inspired others to join. This is an example of how individual acts of integrity can affect the future of entire communities.

VI

The integrity of a few can change the life outcomes of a larger group. This is shown in Akira Kurosawa's 1964 movie *Shichinin no Samurai* (*Seven Samurai*) one of the most famous movies

---

2012]; [online] Available at: http://www.myvidster.com/video/7190041/Police_Brutality_Mother_Tasered_While_Kids_Watch_for_Using_a_Cellphone [Accessed on August 21 2012]; [online] Available at: http://mg.co.za/article/2012-03-13-suspected-police-brutality-in-vaalwater [Accessed on August 21 2012].

ever made. It was remade as the 1960 western *The Magnificent Seven* with director John Sturges and has inspired several movies since then. Akira Kurosawa's story rests on the Bushido warrior code of honor and the duty of the warriors. Many samurai were unemployed because the old traditional order was changing but the need for them still existed and their roles had not yet transitioned into the armies or police forces of modernity. Some villagers and a younger samurai observe the master samurai Kambei shave his head to pose as a monk (thus dishonoring his warrior rank) in order to rescue a child that was taken hostage. As a masterless and hungry samurai (ronin) he ignores one aspect of the Bushido code in order to win the battle for the child. The villagers are impressed and ask him to defend their terrorized village and crops from bandits who repeatedly attack them. In essence they are peasants living the kind of unpretentious life that Grey claims is the true human reality. However they cannot defend themselves from the bandits because none of them have the kind of integrity that will allow them to fight together for the good of the community. Instead they are defeated each crop season when they try to protect only their own farms and crops. This is not necessarily an indictment of the villagers because some social insect communities have specialized soldiers and other animal groups have dominant males that protect the group. This depiction of peasants as having less initiative than the warriors (or city dwellers) despite their daily contact with and mastery over nature is also present in European and Catholic traditions (Binde, 2001).

MacGregor (2004) has written about 'survival' or 'subsistence' morality or the moral insight that comes out of so-called unmediated experiences of survival that these villagers would have had. MacGregor (2004) correctly observes that people exhibiting a 'subsistence' or 'barefoot epistemology' do not choose their conditions and 'lifestyle' does not necessarily determine human morality. If this village had existed in the 1990s my alma mater would have used actor-oriented social theory in order to understand why the villagers did not work together to defeat the bandits. Röling (2001) pointed out that this actor-based approach detracts from the social contract of the social sciences because it focuses on the reasons why people make selfish choices in social dilemmas while neglecting the conditions under which people make co-operative choices, and it also dissuades graduate students from undertaking praxis research that would lead to sustainable futures. Röling's comments fit with the recognition of the community-centred praxis approach that all action, even the casual observation characteristic of positivistic research affects a system and that inaction is consequential and thus ultimately partisan (Warry, 1992; Singer, 1994; Nereu et al., 1997).

Kambei and the younger samurai Katsushiro agree to fight for food. I disagree with Keough (2008) that their Bushido code of conduct was ignored and that food was the primary motivator. The task fell within their Samurai lifestyle and could lead to future jobs. Other ronin are recruited. A rejected ronin follows them to the village and uses a trick on the villagers in order to become accepted as the seventh. The samurai strengthen the village defenses and teach the villagers to fight. They also discover that the villagers have killed less moral samurai-bandits in the past. Some of the outlying farms have to be sacrificed in order to save those in the core. It was the integrity and skills of the samurai that convinced those unlucky villagers to sacrifice their homes. In the final battle only three warriors survive and Kambei says to Shichiroji his old friend that 'once again we've survived ...We have lost. So. Again we are defeated. The farmers have won. Not us..'

Kambei is implying that the farmers who continue in their old ways have lost little and retain their purposeful lives, while the samurai live only until the next battle. However by maintaining a warrior code that allows them to live lives that suit their natures the Samurai can also benefit society (Keough, 2008). As stated by Robert Papazian ... 'it's the effect of one's life and death on others that counts.' These individual acts of integrity can inspire incremental or even major change that allows societies to evolve in a direction that make all lives more worthwhile.

In the children's movie version of the *Seven Samurai* (Pixar Studio's *A Bug's Life*), there is a different kind of tension between the individual and the group and a different presentation of integrity. Instead of the Bushido code, the film audience is introduced to the artistic integrity of the circus group of bugs (the warriors), Aesop's grasshopper becomes a bandit group extracting food tributes (surplus value?) and the integrity of an inventor ant Flik who is different from the collective. This ant society has a queen and two princesses but no soldier ants. Flik creates his inventions outside the normative constraints of society in the hope of providing a societal benefit (to find oneself in ethical praxis is thus to find oneself with a job to do - Lysaker, 2011). However his untested inventions cause problems. It's also noteworthy that the younger princess Dot helps Flik in a way that does not imply sibling rivalry with Atta who is the romantic interest of Flik.

VII

Philosophers of science study various fields to determine the factors or problems underlying the present conduct of science, studying whether scientists are making correct decisions and then using their conclusions to make normative recommendations as a basis for the initiating, modifying, ending or for the future conduct of science (Pinnick and Gale, 2000). Academic integrity includes reason, scholarship, community, diversity, dignity, intellect and respect (Cox et al., 2012). In Samuel Johnson's novel *Rasselas, Prince of Abyssinia*, his character Imlac was told that: 'Integrity without knowledge is weak and useless, and knowledge without integrity is dangerous and dreadful'. Scientists need to build community networks and cannot continue to act as if their work should not be directed towards any other goal beyond the need to know the answer to their next experiment; or to finding a model so interesting that repeated testing of it without any further creative thought wins scientific immortality in the form of a lower-case version of the scientist's name attached to an aspect of nature (Pollack, 1992).

VIII

The last problem focused on in this essay is why Higher Education (teaching, research and service) has not been effectively harnessed to achieve the UN Millennium Development Goals (MDGs). This can be considered a reformulation of one of the questions of Doring (1895) - how should society be constituted in order to render the moral will possible to all? These MDG goals cannot be achieved without direct involvement in the governance of the world's poorest countries. The Millennium Development Goals are outlined, their unachieved status is explicated and possibly ways to attain them are suggested through the use of praxis, applied philosophical approaches and instrumentality (Cernea, 1995; Warry, 1992; Loftus, 2009).

Praxis or knowledge-based action can be defined as a property of individuals that emerges from the interactions of the theories (beliefs) that they hold, the actions that they practice, the values that they assume, and the contexts in which they interpret of their society (Kay, 1994; Holland and Ramazanoglu, 1994). Praxis research requires non-alienating methodologies that are dialogic and participatory (Warry, 1992). Applied research typically means analyses of particular human problems, situations, or processes for the purposes of understanding their causes, dynamics, and consequences; and, in some instances, for developing courses of action designed to influence those situations or processes (Rappaport, 1993; Giarelli, 1996). This definition of praxis is similar to that written by Aristotle in his *Nicomachean Ethics* (Lysaker, 2011).

Choosing applied research may bring conflict with community power groups and brokers and also a loss of professional peer support (Cox, 1997; Hastrup and Elsass, 1990; Shore and Wright, 1996). While philosophy is concerned with the meaning of human life, feminists try to add equality and to change patriarchal life rules (Turksma, 2001). There has been a long battle against praxis approaches in philosophy. The split between theory and practice is said to be linked to masculine ideas of science which exclude advocacy, activism and commitment from real, publishable science (Hatten et al., 2000). For example Perry (1920) writes that 'pragmatists and instrumentalists have dressed the intellect in livery and sent it to live in the servants' quarters.'

According to Francis Bacon:

> Those who have handled the sciences have been either Empricists or Rationalists. Empiricists, like ants, merely collect things and use them. The Rationalists, like spiders, spin webs out of themselves. The middle way is that of the bee, which gathers its material from the flowers of the garden and field, but then transforms and digests it by a power of its own. (Bacon, 1620)

Marx addressed these philosophical concerns in his *Theses on Feuerbach*. He develops a theoretical argument that 'philosophy and practice are part of a differentiated unity: learning about the world is a process of changing it, based on the lessons gained from thinking about it and so on' (Lysaker, 2011). The most demanding principle of scientific integrity is to exercise sound citizenship (Holton, 2005). Science can claim moral authority only when its activities honor truth and public interest. Van Steenbergen (1983) and Verijken (1992) discuss the role of science as a 'pull model' in facilitating change.

The first five Millennium Development Goals (MDGs) are:

(1) to eradicate extreme poverty and hunger,

(2) to achieve universal primary education,

(3) to promote gender equality and empower women,

(4) to reduce child mortality and,

(5) to improve maternal health

These five MDGs are the essence of democracy but have only been achieved in Scandinavian countries and socialist economies like Cuba. Venezuela's Hugo Chavez is using his country's oil wealth to achieve some of these goals and other developing countries are working towards them with varying degrees of success.

Canada achieved equality in educational attainment, health and survival. But women scored lower in economic participation and opportunity and on political empowerment than in previous years, with women making up only 21 per cent of MPs and 23 per cent of cabinet posts in Parliament (47th in rankings)[10]. African countries have made the least progress towards MDG5. Almost half of global maternal deaths occur in Africa and this is said to be both a reflection of the poor gender equality in those countries and the influence of US social conservatives on aid policies (Simwaka et al., 2005). Indigenous knowledge could provide cost-savings and remove reproductive health from the prying eyes of the Christian Right (Abrahams et al., 2002).

Goal (8) 'to develop a global partnership for development' is a lofty one. Awarding Nobel Prizes to development practitioners and activists such as Muhammad Yunus (Peace) and Amartya Sen (Economics) can be seen as an initial step towards this goal. The World Bank plan for insecticide-treated bed nets is a low-cost and simple solution to addressing malaria. The medicinal plant Artemisia is also being used to treat malaria and both of these approaches fall into the environmental sustainability mandate of Millennium Development Goal 7[11]. However solutions of this type remain in the minority because 'prestigious research' is rarely developmentally oriented. The epidemiologist Lonnie King (2004) claims that many scientists do not want to deal with policy issues and politics, but without this engagement there can be no improvement in public health.

---

[10] Worl Economic Forum 2007 [online] Available at: www.weforum.org/issues/global-gender-gap [Accessed on August 21 2012].

[11] [online] Available at:www.unmillenniumproject.org/.../malaria-complete-lowres.pdf [Accessed on August 21 2012].

The short-term way for Higher Education (teaching, research and service) to achieve the first five UN Millennium Development Goals before 2015 is for aid money to be spent on salary support for school teachers, doctors and nurses (Doull and Campbell, 2008).

There are many problems with the status quo in development aid. Many researchers consider that food aid undermines local agriculture even though it is justified in emergencies (Del Ninno et al., 2007). It has been argued that aid props up undemocratic regimes (Kono and Montinola, 2009). Only Norway, Denmark, the Netherlands and the United Kingdom gave any string-free aid (Deen, 2004). Eighty percent of aid dollars given by the US has to be spent on American products such as Caterpillar and John Deere tractors. Anti-AIDS drugs from the United States which are forty-three times as expensive have to be bought with American aid money instead of cheaper generic products from South Africa, India or Brazil. Sweden gave 0.83% of their GDP in aid in 2002, 4% of this went to basic education, 0.56% to basic health and less than 2% on water and sanitation (Economic Commission for Africa, 2004).

Africa had 24% of the global burden of disease[12] but only 3% of the health care workforce and 1% of the world's financial resources. The WHO estimates that it will take an additional 2.4 million physicians, nurses, and midwives to meet the needs of the continent, especially of sub-Saharan Africa, together with an additional 1.9 million pharmacists, health aides, technicians, and other necessary personnel. The WHO projects that if all training were to be completed by 2015, it would cost an average of $136 million per country per year. Kumar (2007) reported on efforts by Ghana to pay doctors an allowance. However it was unsustainable for the country and it was resented by nurses who received less. It may be possible for aid agencies to become involved in efforts to provide low cost housing for medical personnel, a program tried in Ghana, if it was part of a research program to test solar housing or green technology housing in developing countries for example.

Worker numbers and quality are positively associated with the survival of infants, children and their mothers (op. cit.). Shortfalls in medical staff are greatest in sub-Saharan Africa and in South-East Asia (because of its large population). 'Poverty, imperfect private labor markets, lack of public funds, bureaucratic red tape and political interference produce [a] paradox of

---

[12] The World Health Report 2006 [online] Available at: www.who.int/whr/**2006**/en/

shortages in the midst of underutilized talent' (unemployed health professionals). Using aid money to provide an honorarium to boost doctors, nurses and teachers' salaries across the African continent could alleviate these problems. In the education sector the money could be given to schools with a 50:50 ratio of girls and boys to help support gender equity. Proper safeguards would have to be put in place to avoid fraud, like having lists of the teachers working in each school before the start of the program. Embassy staff or aid staff may have to pay teachers individually or set up systems to pay the funds directly to their bank accounts to help accountability. Even though the money will go to salaries and not equipment or buildings, teachers could be paid in subsequent years based on the literacy of their pupils. This plan may alleviate the brain drain of medical professionals from Africa, comfort conservative Europeans who do not want immigration, and would indirectly support more than one MDG.

A 2006 WHO study on the lost investment due to the brain drain from Kenya claimed that each individual migrant doctor cost US$517,931 while each migrant nurse cost $338,868[13]. There will always be trained professionals who want to migrate in order to use the most recent equipment and conduct the most topical research; but if their research in Africa was rewarded with publications and prizes, this might change.

Developed world researchers could also be encouraged to do more poverty-alleviating research with colleagues in developing countries. For example –technology using nets to harvest sea spray, reversing desertification and increasing solar technology

Green Revolution research won Norman Ernest Borlaug a Nobel Peace Prize in 1970[14].

Vandana Shiva has claimed repeatedly that the Green Revolution was not a success and that it has led to genetic erosion, pesticide-linked health hazards for farmers and environmental harm. It also had very little impact on poverty alleviation; otherwise the Millennium Development Goal #1 would have been achieved. It is hard to argue with the contention that genetic engineering of seeds brings the food supply and life itself into the power and control of a few multinational companies. At least 166,000 farmers have killed themselves since 1997 because of Green

---

[13] [online] Available at: www.who.int/social_determinants/resources/gkn_packer_al.pdf
[Accessed on August 21 2012].

[14] [online] Available at:
http://www.worldfoodprize.org/en/dr_norman_e_borlaug/short_biography/ [Accessed on
August 21 2012].

Revolution technology (Ramesh, 2008). Farmers have to buy genetically-altered seed, expensive fertilizers and pesticides annually. They take out loans to use this technology but then cannot sell their produce in the globalized market place for enough money to repay their loans.

Development practitioners achieve recognition through the Right Livelihood Award (the Alternative Nobel Prize).Vandana Shiva won this award in 1993. At her lecture at the University of Davis, California on April 25 2006, which I attended, Dr. Shiva stated that she could not work in sustainable agriculture, fighting against GMOs and helping the poor within academia. So she formed her own institute and sought academic collaborators. She made a similar claim in her 1993 Right Livelihood Award Acceptance speech:

> In 1982, I left an academic career with a dream to build an independent research initiative for generating a different kind of knowledge, which would serve the powerless not the powerful, which would not get all its cue from Western Universities and international institutions, but would also be open to learn from the indigenous knowledge of local communities, which would break down the artificial divide between experts and non-experts and subject and object. (Shiva, 1993)

Professor Yunus also stopped teaching what he called 'elegant economic theories' in order to start his microcredit bank that worked with famine victims and the poorest of the poor women (Yunus and Jolis, 2001). Although these are just two examples it indicates that in some parts of the world there are constraints to performing academic work that actually and visibly achieves UN Millennium Development Goals.

The agency and power of normal science lies in the interlocking interests that make up the 'Old School Tie' networks of relationships that exist in academia (Harvard, Yale, Chicago, Cornell, Stanford and Princeton) and in the major global institutions such as the UN agencies, IMF, Goldman Sachs and the World Bank. These relationships are also strong in France and Britain (Calmand et al., 2009). Most of the current British cabinet went to either Cambridge or Oxford as did many of the UK's law lords and top journalists. Before Oxbridge they attended 7% of the country's private, fee-paying schools (Cadwalladr, 2008). Similarly trained people have similar viewpoints of the world; and to question them is heresy. Cornell University researchers Herring and Roberts (2006) remarked on the 'massive protests by indigenous peoples and shantytown residents in Bolivia, Ecuador, and Argentina [who] have recently overthrown governments that implemented the so-called 'Washington Consensus' for free market or neo-liberal reform, a model of development that is widely embraced by economists and powerful international financial institutions'. Structural adjustment policies also weakened African states and increased poverty (Adejumobi, 2006; Stiglitz, 2000; Hanock, 1989).

Actors in long-standing interlocking relationships are sometimes resistant to new ideas that require a loss of, or at least re-negotiation of existing scientific networks and interests. One dominant world view developed out of the 1989 National Institutes of Health / Rockefeller University conference on emerging viruses (King, 2002). The anxieties about HIV, Ebola and Hantaviruses expressed at that conference hardened into an orthodox set of predictions and recommendations that then dominated how Americans funded research into vaccines and then marketed them to developing countries.

There are many examples of the interlocking interests at the apex of science. Mitra (1998) revealed that three of the past five Masters of Trinity College, Cambridge have won the Nobel Prize and that Professor Sen is the 31st Nobel laureate of the university. Does the recognition of genius predict who works at Cambridge, or does a job at Cambridge facilitate the receiving of prizes? (Sen took a 65% pay cut for his Cambridge job). Erren (2008) argued that an open-door research environment with opportunities for interdisciplinary collaboration provides strong incentives to avoid the dead ends in science and focus instead on the solutions of problems of 'paramount, rather than tangential, importance'; it is this nurturing environment that accounts for Cambridge's cluster of Nobel laureates. Nurturing environments for women laureates existed at Hunter College, New York and Washington University, St. Louis. However Lawrence (2007) argued that junior scientists need to publish in the top journals, are cautioned against straying too far from the existing scientific norms or vogues; and should network with potential reviewers of their articles. While senior scientists cling to established theories and hierarchies, preferring to ignore evidence rather than risk making scientific errors (Charlton and, Andras, 2008; Molnar et al., 1992; Sousa and Busch, 1998).

If the ethical challenges were removed from the receiving countries by supplementing the pay for doctors, nurse and teachers directly from abroad with aid money this would reduce corruption. Adejumobi (2006) refers to the patron-client network in which little accountability to the supposed aid recipients exists. He calls for more democracy and participatory participation in order to reduce this source of corruption, while I advocate for less participation in the short term because participation takes time and money. Hyden (2007) quotes research stating that most African countries use networks of kinship loyalties with lower transaction costs first and are less loyal to larger group or nation states. This reinforces the local patron-client relationship that undermines larger-scale aid delivery. This is not only a problem in Africa. The increase in patron-client relationships in the West; for example the growth in conservative think tanks that provide comfortable incomes for scientists despite the veracity of their statements, is considered to reduce the integrity of science and economics (Vidal, 2011; Union of Concerned Scientists, 2004). If the world economy transforms so that it consumes a large quantity of African goods that can be produced sustainably using fair trade principles practiced by a large section of the population (Rodney, 1974; Hyden, 2007) then local participation and institution building would become practical.

If university study and economic migration opportunities were restricted to students and professionals from countries with good gender equality this would put pressure on leaders who wanted to uphold gender inequality. This strategy is similar to that of Nobel laureate P.W. Bridgman who announced in a 1939 *Science* publication that he would close his lab to visiting scientists from totalitarian countries because they had subordinated their loyalty to science to local politics (Holton, 2005). Universal free primary education (for the first three children) and universal low-cost or free health care can reduce poverty and improve child mortality and maternal health. The achievement of other MDGs will have to come from improving the ways that science is undertaken, rewarded and funded.

Policy entails going from what is to what should be. The norms, values and motivations of scientists and their institutions limit their ability to produce low-external input technology (King, 2004). Local sustainable technologies can only come from local practitioners in an ideal world, with a diversity of scientific norms. Or the western scientific norms could be transformed with academic awards that truly reward poverty alleviation. A prestigious UN Millennium

Development Prize could be awarded in each of the relevant categories (1) development economics (2) education delivery (3) gender equality and empowerment (4) basic, appropriate and equitable health care (5) basic, appropriate and equitable maternal care (6) low cost, appropriate and environmentally sustainable treatments for HIV/AIDS and neglected diseases (7) economic environmental sustainability that includes non-human species and an eighth prize for engineering (small-scale appropriate technology). The prize committee could consist of academics with expertise in praxis and the local staff of front-line development NGOs. The awarded research would have to show demonstrable results like the work of Professor Yunus. If the funding can be found to make these prizes worth more than the Nobel Prizes, that is just what skilled applied scientists addressing poverty and inequity deserve.

References

Abrahams N, Jewkes R, Mvo Z. (2002) 'Indigenous healing practices and self-medication amongst pregnant women in Cape Town, South Africa' *Afr J Reprod Health* 6(2):79-86.

Adejumobi Said. (2006) Governance and poverty reduction in Africa: A critique of the Poverty Reduction Strategy Papers (PRSPs). Presented to the Inter-Regional Conference on Social Policy and Welfare Regimes in Comparative Perspectives, University of Texas in Austin, April 20-22, 2006.

Bacon, F. ([1620] 1994) *Novum Organum*. Peter Urbach and John Gibson, trans. and eds. Chicago: Open Court.

Beauchamp, Tom L. (1999) 'The failure of theories of personhood' *Kennedy Institute of Ethics Journal* 9 (4): 309-324.

Bekoff Marc and Pierce Jessica. (2009) *Wild Justice: The Moral Lives of Animals*. Chicago: University of Chicago Press.

Benhabib, S. (1987) 'Rhetorical affects and critical intentions: A response to Ben Gregg' *Theory and Society* 16 (1): 153-158.

Bertram, C. (1997) 'Political Justification, Theoretical Complexity, and Democratic Community' *Ethics* 107 (4): 563-583.

Binde, P. (2001) 'Nature in Roman Catholic tradition' *Anthropological Quarterly* 74: 15-27.

Blaydes, Lisa and Linzer, Drew A. (2008) 'The political economy of women's support for fundamentalist Islam' *World Politics* 60(4): 576-609.

Bracton, H. (1968) *On the laws and customs of England*, Vol. II, Harvard University Press, Cambridge, MA.

Bradshaw G.A., Schore A.N., Brown J.L., Poole J.H., Moss C.J. (2005) 'Elephant breakdown' *Nature* 433(7028):807.

Brosnan, S. F., and de Waal, F. B. M. (2003) 'Monkeys reject unequal pay' *Nature* 425: 297-299.

Cadwalladr, C. (2008) 'Oxbridge blues' Comment. *Guardian Unlimited* 16 March 2008 viewed 9 August 2012, http://commentisfree.guardian.co.uk/carole_cadwalladr/2008/03/oxbridge_blues.html

Calmand J., Giret J-F., Guégnard C., and Paul J-J. (2009) 'Why Grande Ecoles are so valued?' Intl. Conf. DECOWE: Development of Competencies in the World of Work and Education, Ljubljana, Slovenia, Sept. 2009.

Cernea, M.M. (1995) 'Social organization and development anthropology' Malinowski Award Lecture. *Human Organization* 54 (3): 340 - 352.

Charlton, B., Andras, P. (2008) ''Down-shifting' among top UK scientists? – The decline of 'revolutionary science' and the rise of 'normal science' in the UK compared with the USA' *Medical Hypotheses* 70: 465 – 472.

Cohen, G.A. (2011) Rescuing conservatism: A defense of existing value. In: Samuel Freeman, Rahul Kumar, and R. Jay Wallace, eds. 2011. *Reasons and recognition essays on the philosophy of T.M. Scanlon* New York: Oxford University Press, pp. 203 – 230.

Colarelli, S. (2002) 'Conservatives are liberal, and liberals are conservative – on the environment' *The Independent Review* VII (1): 103-107.

Cox, Damian, La Caze, Marguerite and Levine, Michael, "Integrity", *The Stanford Encyclopedia of Philosophy (Spring 2012 Edition)*, Edward N. Zalta (ed.), URL = <http://plato.stanford.edu/archives/spr2012/entries/integrity/>.

Cox, H. (1997) 'Professional responsibility to the communities in which they work and live' *Human Organization* 56 (4): 490 - 492.

DeBrabander, F. 2012. Deluded individualism. http://opinionator.blogs.nytimes.com/2012/08/18/deluded-individualism/?src=me&ref=general. [Accessed on August 21 2012].

Deichmann, Ute, Müller-Hill, B. (1998) 'The fraud of Abderhalden's enzymes' *Nature* 393 (6681): 109 - 111.

Deen, T. (2004) 'Development: Tied aid strangling nations, says U.N.' Jul 7 2004 viewed August 11 2012, http://ipsnews.net/interna.asp?idnews=24509

Del Ninno, C., Doeresh, P. A. and Subbarao, K. (2007) 'Food aid, domestic policy and food security: contrasting experiences from South Asia and sub-Saharan Africa' *Food Policy* 32 (4): 413–35.

Dodd M.D., Balzer, A., Jacobs, C.M., Gruszczynski, M.W., Smith, K.B., Hibbing, J.R. (2012) 'The political left rolls with the good and the political right confronts the bad: connecting physiology and cognition to preferences' *Philos Trans R Soc Lond B Biol Sci.* 367(1589):640-9.

Doring, A. (1895) 'The motives to moral conduct' *International Journal of Ethics* 5 (3): 361-375.

Doull, L., Campbell, F. (2008) 'Human resources for health in fragile states' *Lancet* 371(9613):626-7.

Economic Commission for Africa (2004) *Economic report on Africa 2004: Unlocking Africa's trade potential in the global economy overview*, Twenty-third meeting of the Committee of Experts of the Conference of African Ministers of Finance, Planning and Economic Development, April 28, 2004. Kampala: Uganda. http://www.uneca.org/cfm/2004/overview.htm. [Accessed on August 21 2012].

Elliott, Herschel and Lamm, R. (2002), A moral code for a finite world, *Chronicle of Higher Education*, November 15, 2002, pp. B7-B9. Available at: http://chronicle.com/ free/v49/i12/12b00701.htm

Ericson, M., Park, H., Parlapiano, A., Willis, D. (2012) 'Who's financing the 'Super PACs'' 31 Jan 2012, viewed 9 August 2012, http://www.nytimes.com/interactive/2012/01/31/us/politics/super-pac-donors.html

Erren, T. (2008) 'Hamming's "open doors" and group creativity as keys to scientific excellence: The example of Cambridge' *Medical Hypotheses* 70: 473 – 477.

Eubanks, V. (2009) 'Double-bound: Putting the power back in participatory research' *Frontiers: A Journal of Women's Studies* 30 (1): 107-137.

Flyvbjerg, B. (1998) 'Habermas and Foucault: Thinkers for civil society?' *British Journal of Sociology* 49 (2): 210-233.

Farazmand, A. (1999) 'Privatization or reform? Public enterprise management in transition' *International Review of Administrative Sciences* 65: 551-56.

Foucault, M. (1980) *Power/Knowledge: Selected interviews and other writings 1972-77*, translated by C. Gordon. New York: Pantheon Books.

Giarelli, G. (1996) 'Broadening the debate: The Tharaka participatory action research project' *Indigenous Knowledge and Development Monitor* 4 (2): 19 - 22.

Goldman, M., Schurman, R.A. (2000) 'Closing the "great divide": New social theory on society and nature' *Annual Review of Sociology*, 26: 563-584.

Gray J. (2002) *Straw dogs: Thoughts on humans and other animals.* London: Granta Publications.

Gray J. (2004) *Heresies: Against progress and other illusions.* London: Granta Publications.

Griffiths, John G. (1980) *The origins of osiris and his cult.* Leiden, Netherlands: E. J. Brill.

Hanock, G. (1989) *Lords of poverty: The power, prestige, and corruption of the international aid business.* New York: The Atlantic Monthly Press.

Hastrup, K., Elsass, P. (1990) 'Anthropological advocacy' *Current Anthropology* 31 (3): 301 - 311.

Hatten, R., Knapp, D. and Salonga, R. (2000) 'Action research: Comparison with the concepts of 'the reflective practitioner' and 'quality assurance'. Action Research E-Reports, 8. Available at: http://www.cchs.usyd.edu.au/arow/arer/008.htm. First published, 1997.

Herring, R., Roberts, K.M. (2006) *Contentious politics: Science, social science and social protest.* [Theme proposal] Institute for Social Sciences, Cornell University.

Holland, J., Ramazanoglu, C. (1994) Coming to conclusions: power and interpretation in researching young women's sexuality. In: M. Maynard, J. Purvis,ed. 1994. *Researching women's lives from a feminist perspective.* Taylor and Francis, London, pp. 125 - 148.

Holton, G. (2005) 'Candor and integrity in science' *Synthese* 145 (2): 277-294.

Huang, C. (2011) 'Facebook and Twitter key to Arab Spring uprisings: Report' *The National.* June 6 2011, viewed August 22, 2012.

Hyden, G. (2007) 'Governance and poverty reduction in Africa' *PNAS* 104 (43): 16751-16756.

Johnson, S. (1909) *Rasselas, prince of Abyssinia.* London: Cassell & Company.

Kahan, Dan M., Peters, Ellen, Wittlin Maggie, Slovic Paul, Larrimore Ouellette Lisa, Braman Donald, Mandel G. (2012) 'The polarizing impact of science literacy and numeracy on perceived climate change risks' *Nature Climate Change*, 2012; DOI: 10.1038/NCLIMATE1547

Kanai R, Feilden T, Firth C, Rees G. (2011) 'Political orientations are correlated with brain structure in young adults' *Curr Biol.* 21(8):677-80.

Kay, J.W. (1994) 'Politics without human nature? Reconstructing a common humanity' *Hypatia* 9 (1): 21 - 52.

Keough, K. (2008) 'Cowboys and shoguns: The american western, japanese jidaigeki, and cross-cultural exchange' *Senior Honors Projects.* Paper 106. The University of Rhode Island. Online at http://digitalcommons.uri.edu/srhonorsprog/106

King, N. (2002) 'Security, disease, commerce: Ideologies of postcolonial global health' *Social Studies of Science* 32: 763 – 789.

King, L. (2004) 'Impacting policy through science and education' *Preventive Veterinary Medicine* 62: 185 – 192.

Kirigia, J.M., Gbary, A.R., Muthuri, L.K., Nyoni, J., Seddoh, A. (2006) 'The cost of health professionals' brain drain in Kenya' *BMC Health Serv Res.* 6:89.

Kono, Daniel and Montinola, G. (2009) 'Does foreign aid support autocrats, democrats, or both?' *The Journal of Politics* 71: 704-718.

Kumar, P. (2007) 'Providing the providers - remedying Africa's shortage of health care workers' *N Engl J Med.* 356(25):2564-7.

Lakoff, George, Wehling E. (2012) *The little blue book: The essential guide to thinking and talking democratic.* New York: Simon & Schuster.

Lawrence PA. (2007) 'The mismeasurement of science' *Curr Biol* 17: R583–585.

Lawrence, D. (2010) How political activists are making the most of social media, Forbes 15 Jul 2010, viewed August 12, 2012 http://www.forbes.com/2010/07/15/social-media-social-activism-facebook-twitter-leadership-citizenship-burson.html

Loftus, A. (2009) 'The theses on Feuerbach as a political ecology of the possible' *Area* 41: 157–166.

Lysaker, J, (2011) 'Praxis and form: Thirty notes for an ethics of the future' *The Journal of Speculative Philosophy* 25 (2): 213-238.

Lunn-Rockliffe, S. (2011) Christianity and the Roman Empire. www.bbc.co.uk/.../romans/christianityromanempire_article_01.shtml [Accessed August 22, 2012].

Mason, J.J. (2000) 'Room at the top for women? - Nobel Prize women in science by Sharon Bertsch McGrayne' *Trends in Genetics* 16: 96 – 97.

Mayer, J. (2010) 'Covert operations. The billionaire brothers who are waging a war against Obama' *The New Yorker* August 30, 2010, viewed on August 22, 2012.

http://www.newyorker.com/reporting/2010/08/30/100830fa_fact_mayer

Merchant, C. (2006) 'The scientific revolution and the death of nature' *Isis* 97 (3): 513-533.

Mitra, S. (1998) Amartya Sen the conscience of economics. www.india-today.com/itoday/26101998/cover.html

Moffett, M. W. and Tobin, J. E. (1991) 'Physical castes in ant workers: a problem for *Daceton armigerum* and other ants' *Psyche* 98: 283–292.

Molnar, J.J., Duffy, P.A., Cummins, K.A., Van Santen, E. (1992) 'Agricultural science and agricultural counterculture: Paradigms in search of a future' *Rural Sociology* 57: 83 - 91.

Perry, R. (1920) 'The integrity of the intellect' *The Harvard Theological Review* 13 (3): 220-235.

Pinnick, Cassandra , Gale, G. (2000) 'Philosophy of science and history of science: A troubling interaction' *Journal for General Philosophy of Science* 31 (1):109-125.

Pollack R. (1997) 'A crisis in scientific morale' *Nature* 385(6618):673-4.

Ramesh, R. (2008) 'India pledges £7.6bn to combat rural suicides' *The Guardian* March 1 2008, viewed on August 1, 2012. http://www.guardian.co.uk/world/2008/mar/01/india.

Rappaport, R.A. (1993) 'Distinguished lecture in general anthropology: The anthropology of trouble' *American Anthropologist* 95 (2): 295 - 303.

Rodney, W. (1974) *How Europe underdeveloped Africa*. Howard University Press.

Roling N. (2001) From arena to interaction: Blind spot in actor-oriented sociology. admin_en_Roling_long_conference_(2).pdf

Ross, M. (2008) 'Oil, Islam, and women' *American Political Science Review* 102 (1): 107 – 123.

Scott, A. (2007) 'Peer review and the relevance of science' *Futures* 39: 827 – 845.

Schmidtz, D. (1997) 'When preservationism doesn't preserve' *Environmental Values* 6: 327–39.

Shiva, V. (1993. 'Diversity and freedom' Acceptance Speech by Vandana Shiva December 9th, 1993. Online at http://www.rightlivelihood.org/shiva_speech.html

Shore, C., Wright, S. (1996) 'British anthropology in policy and practice: A review of current work' *Human Organisation* 55 (4): 475 - 479.

Siebert, C. (2006) 'An elephant crackup?' *The New York Times* 8 Oct 2006 viewed on August 22 2012, http://www.nytimes.com/2006/10/08/magazine/08elephant.html?pagewanted=all.

Simwaka BN, Theobald S, Amekudzi YP, Tolhurst R. (2005) 'Meeting millennium development goals 3 and 5 - Gender equality needs to be put on the African agenda' *British Medical Journal* 331(7519):708–709.

Sousa, Ivan de, Busch, L. (1998) 'Networks and agricultural development: The case of soybean production and consumption in Brazil' *Rural Sociology* 63: 349 – 371.

Stiglitz, J. (2000) 'What I learned at the World Economic Crisis' *The New Republic* 17 April 2000, viewed on August 1 2012, Available at http://www2.gsb.columbia.edu/faculty/jstiglitz/download/opeds/What_I_Learned_at_the_World _Economic_Crisis.htm

Thielke, T. (2005) 'For God's Sake, Please Stop the Aid!" Interview with the Kenyan economics expert James Shikwati. Translated from the German by Patrick Kessler http://www.spiegel.de/international/spiegel/0,1518,363663,00.html

Turksma, R. (2001) 'Feminist classic philosophers and the other women' *Economic and Political Weekly* 36 (17): 1413-1417.

Union of Concerned Scientists (2004) *Scientific integrity in policy making* July 2004 http://www.ucsusa.org/assets/documents/scientific_integrity/scientific_integrity_in_policy.

U.S. House of Representatives. Committee on Government Reform – Minority Staff. (2003) *Politics and science in the Bush Administration: Prepared for Rep. Henry A. Waxman* (August 2003). [Accessed on Augus 1, 2012] Available at http://reform.house.gov/min.

van Steenbergen, B. (1983) 'The sociologist as social architect: A new task for macro-sociology?' *Futures* 15: 376 - 386.

Van Willigenburg, T. (2000) 'Moral compromises, moral integrity and the indeterminacy of value rankings' *Ethical Theory and Moral Practice* 3 (4): 385-404.

Vereijken, P. (1992) 'A methodic way to more sustainable farming systems' *Netherlands Journal of Agriculture Science* 40: 209-223.

Vidal, J. (2011) Climate sceptic Willie Soon received $1m from oil companies, papers show, *The Guardian* 28 Jun 2011, viewed on August 12, 2012 http://www.guardian.co.uk/environment/2011/jun/28/climate-change-sceptic-willie-soon.

Warry, W. (1992) 'The eleventh thesis: Applied anthropology as praxis' Human Organization 51 (2): 155 - 163.

Walan, Magnis and Ljungman A. (2004) The reality of aid. Sweden aims for coherent approach, In: The Reality of Aid Management Committee. *An Independent Review of Poverty Reduction*

and Development Assistance. J. Randel, T. German and D. Ewing (eds.) Viewed August 12,
2012. Available online at
http://www.realityofaid.org/userfiles/roareports/ROA_2004_part6.pdf#nameddest=Sweden&
page=108

World Health Organization (2006) *Working together for health: The World Health Report 2006*.
Viewed August 12, 2012, Available online at http://www.who.int/whr/en/

Yunus, M. and Jolis, A. (2001) *Banker to the Poor*, Oxford University Press.

Co-operatives as an alternative model of social organization

Abstract

In this essay I argue that co-operatives offer an alternative model of social organization in the form of the social economy. People and organizations turn to the social economy to overcome some of the core problems within contemporary capitalism such as global gender and gage inequality and unemployment. There are examples in Europe in which co-operatives have altered entire regions from rundown to dynamic. Writers have referred to these regional successes using terms such as transformative and utopian.

Co-operatives as part of the social economy

For most of the world's developing countries, the 1990s were a decade of frustration and disappointment. The economies of sub-Saharan Africa and Latin America did not rebound economically in response to the structural adjustment prescriptions of the World Bank and IMF (Rodrik 2001; 2002). Frustration with the World Bank and IMF led to the development of many co-operatives in Latin America (Miller, 2006).

Involuntary unemployment is capitalism's most costly market failure and the demand for social services like the social-professional reintegration of disadvantaged groups usually cannot be provided solely by national governments (Monzón Campos, 1997). An alternative economy often arises in response to unemployment. This alternative economy is composed of co-operatives and NGOs working on small projects for community economic development and ethical businesses providing services (camps, financing, daycare, media, housing, women's centres) (Corcoran and Wilson, 2010).

Other groups working in the social economy include credit unions, fair trade organizations, women's groups, aboriginal and anti-poverty organizations, non-profits and some trade unions. This alternative economy is differentiated mainly by the types of businesses involved and whether cash is dominant or if barter arrangements are used. These alternative firms could replace many private firms. In fact Hansmann (1999) makes no distinction between a capitalistic firm and a producer or consumer co-operative, writing that the investor-owned business cooperation is nothing more than a lenders' co-operative or a capital co-operative. This definition of a firm is a boon to those who have to debate the point with stubborn others that "there is no alternative ("Tina") to capitalism (Wolff, 2012).

Whether the alternative economy could become dominant depends on how successfully these organizations could integrate horizontally, how strong their relationships of mutual aid and exchange are and if they could provide representatives to lead regional and national governments. Success in building these networks has been seen in Brazil, Spain, Argentina, Columbia and Venezuela and these networks of co-operatives have proved transformative for poor people and are not mere visions of future utopian societies (Miller, 2006).

Proof that alternative economies are not utopian fantasies can be seen in Europe (Corcoran and Wilson, 2010). The Emilia Romagna region in Italy owes its prosperity, low inequality, high social cohesion and high social capital to the 6% of its workforce that are involved in worker co-operatives. In Italy Article 45 of the Constitution recognizes the social function of co-operation.

Spain also recognizes co-operation in its Constitution –
Article 129[15] [Participation, Cooperatives]
(1) The law shall establish the forms of participation of those interested in Social Security and in the activities of the public agencies whose function directly affects the quality of life or general welfare.
(2) The public authorities shall effectively promote the various forms of participation in enterprise and facilitate cooperative enterprises by means of appropriate legislation. They shall also establish the means that will facilitate access by the workers to ownership of the means of production (Corcoran and Wilson, 2010).

---

[15] http://www.servat.unibe.ch/icl/sp00000_.html

In France, the model of a multi-stakeholder co-op or *Société coopérative d'intérêt collectif* (SCIC) was adapted from Italy (Corcoran and Wilson, 2010). A SCIC acts at the regional level to promote local development projects that look after the public interest in collaboration with local authorities and other partners.

The social economy was originally associated with utopianism and social change. Under the social change framework co-ops could provide services relinquished by the state while acting as agents of economic transformation and community resilience. The social economy paradigm was adopted by the Catholic Church, and by certain European governments (French and Belgian) and then by the EU in the 1980s (Moulaert and Ailenie, 2005; Côté and Fournier, 2005; Fontan and Shragge, 2000). The concept was revived by French academics Henri Desroches, Michel Rocard, Charles Gide and Léon Walras (Laville et al. 2008). One of the reasons for its adoption was for its potential to address the crisis in the welfare state and the negative effects of globalization. Co-operatives were thus seen by some only as another way of organising businesses within the dominant capitalist economy.

Definitions of the social economy

Western Economic Diversification Canada categorizes a social enterprise as a specific business that produces goods and services for the market economy, but manages its operations and directs its surpluses in pursuit of social and economic goals. The social economy is comprised of social enterprises, co-operative development and the third sector (Salkie, 2005). Other definitions are given below:

1. The social economy refers to all initiatives that are not a part of the public economy, nor the traditional private sector, but where capital and the means of production are collective (Neamtan, 2002). The social economy consists of an ensemble of activities and organisations, emerging from collective enterprises that pursue common principles and shared structural elements (Neamtan, 2002):

- The objective of the social economy enterprise is to serve its members or the community, instead of simply striving for financial profit;
- The social economy enterprise is autonomous of the State;
- In its statute and code of conduct, it establishes a democratic decision-making process that implies the necessary participation of users and workers;
- It prioritises people and work over capital in the distribution of revenue and surplus;
- Its activities are based on principles of participation, empowerment, and individual and collective responsibility.

2. Westlund and Westerdahl (1996) articulated three hypotheses on the social economy in Europe. The vacuum hypothesis posits that the social economy can provide support when the public sector shrinks and the private sector does not hire and also shrinks. For example in Russia poor government support is supplemented by informal networks between neighbours, friends and relatives[16] (Sätre Åhlander, 2000). In Québec, co-operatives perform community services (environmental protection, etc.) that are being downloaded by the state (Lewis, 2004). The influence hypothesis assumes that the social economy takes on the roles that the public service sheds through contracts (similar to the social investment state). The local-identity hypothesis states that the social economy grows in the form of local initiatives as a reaction to the negative consequences of globalization. This is what happened in the Emilia Romagna region in Italy where social co-operatives provide health care (Corcoran and Wilson, 2010).

3. Mullan and Cox (2000) define the social economy as "that spectrum of activity located between the public and private sector (and so driven neither by the logic of capital nor by that of the state) which is a form of economic organisation aimed at addressing social need. Social viability and sustainability is placed on a par with economic viability and sustainability with the two being interdependent." "In Ireland, the social economy is represented in nascent form by community-driven efforts to provide essential services which improve the quality of life and to address the gaps in facilities and services which communities have been deprived of but

---

[16] Unfortunately Sätre Åhlander, 2000 does not completely understand what social capital is and links it only to the formal economy.

which are essential in terms of day-to-day living. A large number of social economy enterprises constitute the adding of an economic dimension to work performed which has been historically undervalued, unvalued and unpaid: caring services, maintenance services, cultural activity and community banking. Services can be provided by and for communities on a basis which is more sustainable than simple subsidised service provision."

4. The International Center of Research and Information on the Public, Social and Cooperative Economy (CIRIEC) in Spain has the following definition: "a group of private companies created to meet their members' needs through the market by producing goods and providing services, insurance and finance, where profit distribution and decision-making are not directly linked to the capital contributed by each member, each of whom has one vote. The social economy also includes non-profit organisations that are private non-market producers, not controlled by general government, produce not-for-sale services for specific groups of households and whose main resources come from voluntary contributions by the households as consumers, payments from the government and income from property" (Monzón Campos, 1997; Chaves and Monzón, 2000).

5. An operational definition of the social economy would set boundaries according to the following criteria set out by Arthur et al. (2003):
- Ownership – locally-based and owned largely by its employees (embedded in the community),
- Control – degree of power in decision-making and management control within the enterprise,
- Values – mutualism or reciprocal interdependence, not profit maximisation,
- Product – preferably socially beneficial ,
- Source of finance - majority of value owned by employees or local community; not totally dependent on grants.

6. Sätre Åhlander (2001) looked at the role of the social economy with respect to employment, welfare, rural development and as a model of societal change. She defined

the social economy from a macro-economic perspective as the user-oriented third economic system beside the centrally managed planned economy and the market economy. The social economy then comprises activities for which users take economic decisions. Mutuality or close connections between producers and users are critical for the social economy. This strengthens its territorial characteristic – meaning local jobs and individual services. Also important is the growth of co-ops that have created opportunities for the socially excluded. The focus is therefore on the dynamic process of societal change and the role that the social economy plays in that change.

The Social Investment State as a co-option of the social economy

Federal governments in Europe and Canada have turned to community-based processes and human capital investment to provide local solutions to local problems in increasingly complex and diverse neighbourhoods. This policy has been called the "Third Way" by Anthony Giddens or the "social investment state" (Perkins et al., 2004). David Cameron's "Big Society" fits this framework as well. This policy recognizes that competing in the global economy can take two forms:
(1) A race to the bottom in terms of the welfare of citizens, or
(2) Providing a highly adaptable, skilled and educated workforce that can respond to the 21st century's flexible knowledge-based economy.
The social investment state is a response to neo-liberal critiques of social spending as wasteful and an economic drain. This has led the push to fund only those programs deemed to be more cost effective than welfare, income support and anti-drug and anti-crime spending (Perkins et al., 2004). The social investment state has been criticised for focussing on children as the worker-citizen of the future rather than on retraining adults, for being gender-biased, and for continued adherence to the neo-liberal macroeconomic framework (lack of concern for the environment and commitment to privatization of public service work).

The Irish version of the social investment state "partnershipping" was called "co-optation by the state. By promising limited funding and community consultation in statutory decision-making, the state converted activists into subcontracted civil servants" (Mullan and Cox, 2000). Activist organizations spent their resources on writing funding applications that were better than those of their competing activist organizations and needed to undertake legal incorporation, financial auditing and restructuring to fit EU guidelines. They had less time to spend on understanding structural problems and developing broader alliances, their radical language was not incorporated into funding applications and in most cases local activism did not fit the categories that could be funded. Organizations also needed to hire credentialed grant managers who were not always community members (Mullan and Cox, 2000). Welfare and childcare needs were shortchanged.

Government support for the social economy - What would the world look like?

The social economy in Canada includes at least 175,000 non-profits. There are 78,000 non-profits with charitable status. These generate revenues of $90 billion annually and employ 1.3 million staff members (Quarter et al., 2003). The social economy includes 10,000 co-ops that generate $37 billion a year and employ 150,000 people. More recent figures on the non-profit sector estimate its size at $79.1 billion or 7.8% of the GDP with 2 million people (11% of the workforce) (Geller and Salamon, 2007). Some Canadian provinces are more supportive of the social economy than others. For example Manitoba provided multi-year investment in community capacity. They also created a cabinet committee to examine all government activities through the lens of community economic development. Most childcare spaces in Manitoba are provided by non-profits (Prentice and McCracken, 2004). The Canadian Community Economic Development (CED) Network (CCEDNET) asked the federal government for more investment in the social economy. CCEDNet was founded by 16 CED organizations in 1999 (Lewis, 2004).

In 2002 CCEDNet allied with the *Chantier de l'économie sociale*, Québec's social economy "network of networks". They lobbed federal politicians to create a "Three C" policy environment:

- multi-year funding
- tax credit incentives to mobilize community financial capital

- policy changes and increased funding aimed at solving several serious problems hampering the effective development of human capital

In Quebec, there are an estimated 65,000 people working in 6,200 social economy enterprises which generate annual sales in excess of $4 billion. The social economy in Québec includes all of the co-operatives, mutual benefit societies and associations. Many are linked to the Desjardins credit union movement. Excluding the Desjardins movement, and the two largest agricultural co-operatives, the Québec based social economy has over 10, 000 community organisations with more than 100, 000 workers (Neamtan, 2002). These organizations have developed new organisational methods and new market relations (multiple enterprises and partnerships, fair trade, alternative trade, networking), as well as new types of enterprises with new legal statuses such as social co-operatives or enterprises for social purposes (Neamtan, 2002).

The Federal Government of Canada under the former Liberal government of Paul Martin intended to harness the power of the social economy in Canada. They outlined their strategy in a February 2004 Speech from the Throne; agreeing to support "the efforts of the people who are applying entrepreneurial creativity, not for profit, but rather to enhance the social and environmental conditions in our communities right across Canada." In his opening speech to Parliament, former Prime Minister Paul Martin announced that social enterprises would be assisted by a fund of CAD$152 million in addition to the existing small business programs and business financing programs (ACCORD, 2003). The budget included an expanded mandate for Community Futures Development Corporations. Budget 2004 also provided new funding through pilot programs focused on capacity building ($17 million), financing, generating employment and research ($15 million). The budget would have provided $100 million in credit and "patient capital" for the next five years. The Fiducie in Québec offers "patient" capital funding product with no capital repayment for 15 years, available for real estate or working capital (Favreau, 2008).

In his speech, Martin outlined his background in community economic development (ACCORD, 2003). Martin revealed his history with *Regroupement économique et social du Sud-Ouest* (RESO), a successful collaborative venture between unions, businesses, and community groups in a large rundown area of southwest Montreal. RESO transformed the area into a dynamic part of the city using the tools of community economic development and the social economy. As Regional Development Minister for Québec in the early '90s Martin made multi-year funding available for the core operations of RESO. Martin created a federal government partnership with the Province of Québec and the Québec Solidarity Fund to create a $5 million equity investment pool for RESO development projects (providing the three Cs of CED - capacity building, community capital, and competence (Lewis, 2004).

Other initiatives announced in the Canadian 2004 budget included:
A new Not-for-Profit Corporations Act (S.C. 2009, c. 23), designed to reduce the regulatory burden on the non-profit sector and improve financial accountability,

- Acceptance of most of the recommendations on the tax treatment of charities made by the Joint Regulatory Table (between the government and the voluntary sector), (Voluntary Sector Initiative (Canada), 2003),
- A government investigation into the feasibility of establishing a bank for the charitable sector.

Unfortunately a Conservative government took power in 2006, and not all of the promised activities were undertaken. The funding that had already been transferred such as research funding remained in place (Smith and McKitrick, 2010).

The Social economy and gender

Côté and Fournier (2005) wrote that gender equity became worse under state control. Gender-based approaches and organizations present in the social economy from 1996 – 1999 generated more total jobs, more jobs for women and better paid jobs than after 1996 when the Québec government became involved in regulating and financing the third sector after the Socio-Economic Summit of October 1996 (Status of Women Canada, 2004). Prior to 1999 the social economy was funded by 17 regional funding committees (CRÊS) and women were represented on the management boards alongside provincial government employees. These boards ensured that projects with social ends were funded. The prevalence of women in co-ops is echoed in

Europe where women find co-ops to be a convenient way to organize work, to share responsibility and to have democratic management (Sätre Åhlander (2000). Gender equity worsened when the Québec government instituted a gender-neutral approach from 1999, with the establishment of the *Centre Locaux de Dévelopment* (CLDS) and the *Politique de Soutien au Développment Local et Régional* (Côté and Fournier, 2005). The CLDs were service agencies for small business and only funded groups with the potential to be self-financing after one year; this ironically reduced the number of created jobs, led to fewer jobs for women and fewer well-paid jobs. Funded groups had to use business plans and reporting schemes that were not designed for their organizations and that may have compromised their initial aims. Community initiatives were no longer recognized. This outcome is similar to that described above by Mullan and Cox (2000) in the Irish social investment state. Women were not given a place on the boards of the CLDs even though they were asked to voluntarily train those who took over the CLDs.

Researchers Louise Toupin and Nadine Goudreault evaluated the social profitability of women's work and its contribution to community life so that it can be included in future definitions of the social economy. This measurement is necessary so that women's work can be accorded value in the market-based or currently-defined social economy. Women's groups lost the opportunity to insert their unpaid contributions into the budget allocation criteria in Québec in 1999. Status of Women Canada (2004) funded a study that examined how the restructuring of the social economy negatively impacted the health and working conditions of women in Québec. When Québec shortened hospital stays, women had to bear the burden of the transfer of care delivery to the home and had to perform more complex care. These studies have found that the shift to the social investment state (rather than social economy), in combination with "gender-neutral" policies (which are actually gender-biased) had a negative impact on women.

The Social Economy and a sustainable food system

The traditional definition of food security[17] does not emphasize local production and ensuing community resilience or environmental issues. A proper definition of food security should

---

[17] The World Food Summit (1996) definition of Food Security: "Food security exists when all people, at all times, have physical and economic access to sufficient, safe and nutritious food for a healthy and active life."

include three aspects:

1. Land tenure,

2. Environmental issues as part of sustainable production and

3. Safe food from a consumer point of view.

In Vancouver, British Columbia, a coalition worked on the food security issue. It included the Food Assessment Research Team and the Centre for Sustainable Community Development (CSCD). Their vision of food security includes re-localizing food production, rooftop gardens and urban agriculture, direct sales from farms to institutions, and food related social enterprises. They do not consider their vision to be utopian but rather consider it a viable strategy that should receive financial support from Western Economic Diversification Canada. The community food sector includes community kitchens, community gardens, good neighbour programs, co-op grocery stores, buying clubs and healthy food vending.

Vancouver's Forum of Research Connections (FORC) claims that a social economy approach to addressing acute food insecurity would improve food quality, support local growers and create jobs for current charity recipients.

Agricultural co-operatives provide sustainability. Although they are negatively affected by trade liberalisation and the increasing concentration taking place in the food industry, co-operatives and producer marketing boards empower farmers, allowing them to forge their own paths (Moran et al., 1996; Côté et al., 2000). Historically co-ops have played important roles in the production, processing and sale of foods. Today, many agricultural co-ops are in trouble because of low commodity prices, intensifying competition, high capital needs, growing membership diversity and dwindling government support. They are often vulnerable organisations buffeted by international forces (often financial).

Agricultural co-ops have been significantly affected by trade liberalization and increasing concentration within the food producing and processing industries. Multinational corporations use large-scale corporate agriculture that is increasingly short-lived, mobile and unsustainable; it reduces smallholders to producing cheap, standardized commodities for industrial food processing (Nigh, 1999). Dominant trends in the food business under globalization are: industrialization and subcontracting, global concentration of food sectors, effective and dynamic food distribution, marketing strategies based on product quality, and a uniformity of consumption practices in the international market (Renard, 1998). In Canada many large co-operatives in the grain, dairy and mushroom industries have demutualised.

The Cortes Island Shellfish Growers Co-operative Association is one of the founding members of the new BC Maritime Resource Co-operative which was established to address problems of concentration of power in the industry. They provide shared services to small shellfish co-operatives (oysters and clams). They made large investments in research and development of the shellfish industry in order to support resource-based communities now facing lean times. Environmental problems faced by the industry are bad weather, predators and toxic blooms.

In Canada co-ops are involved in the local provision of organic foods and are building an alternative food economy (for example the multi-stakeholder Growing Circle Food Co-operative on Saltspring Island). The alternative food economy includes Community Supported Agriculture (CSA), food charters, the Slow Food movement, food box programs and various NGOs. Their aim is to create a local and sustainable food system that improves access to fresh and healthy foods and gives producers a greater share of the consumer dollar. These initiatives also fit the frame social economy defined as a grass-roots based, regionally oriented federation of decentralized, autonomous and democratic enterprises building sustainable communities (Wilkinson and Quarter, 1995).

Asia

In Asia there have been strategic alliances of co-ops in production, processing and marketing for tropical fruit. The tropical fruit industry has the classic characteristics that historically have given rise to large and effective co-operative enterprises: i.e., extensive supply, considerable shrinkage because of inadequate technology and weak marketing mechanisms. Malaysian co-operatives work in consumption goods, tourism and agriculture. The social consequences of large-scale production of fruit for export have resulted in loss of land by small holders to plantations owned by multinationals and large-scale growers, transfer of land from local food production to export production resulting in food imports, labour and health/environment issues. An Alternative Trade network has evolved in the last 15 to 20 years to offer producers and consumers a fair and equal relationship, which also provides environmental standards and healthy working conditions. This movement now boasts sales in excess of $200 million worldwide. In Asia tropical fruit (mangosteen, longan and duku) are extensively grown by small holders (over 80%) and in home gardens, and can contribute to income generation. However some fruit yields are low and not all can be commercialized. Asian fruit production and export are addressing issues connected to postharvest, handling, processing and marketing infrastructures. The major issue is that the international market requires standardized high quality products.

Tropical fruits have become important to Southern countries as an alternative to traditional plantation crops like cocoa, coffee, oil palm, rubber and sugar that have low product prices. Consumer interest in organic and healthy foods has provided a stimulus to tropical fruit production. The increase in international tourism means that transportation is available, facilitating the exports of fruit. Between 1987 and 1997 the quantity of fruits such as bananas, grapes, pineapples, citrus and melons imported into Canada from the South increased by approximately 45 percent, from 452 to 656 million kilograms. In 1995 40% of fruit imports came primarily from Latin America. Other significant sources are Morocco and Thailand. The most important fruit imports were bananas, grapes, citrus fruits, melons and pineapples. Other topical fruits include Durion, Guava, Mango, Mangostene, Lanalum, Papaya, Pomelo, Rambutan, Sawo and Starfruit. Mangoes dominate all other crops in terms of production. Other tropical fruit produced at significant levels are pineapples, avocados and papaya (FAO, 2001).

Africa

In Africa co-ops have been making an impact on land tenure and micro-credit issues for the benefit of female farmers. In Nigeria, women make up between 60 and 80 percent of the agricultural labor force, depending on the region, and produce two-thirds of the country's food crops. As elsewhere in Africa, however, extension services had focused on men and their farm production needs and women have difficulties accessing land and credit. When women do own land, the land holding tends to be smaller and located in more marginal areas. Since rural women have less access than men to credit, this limits their ability to purchase seeds, fertilizers and other inputs needed to adopt new farming techniques.

Only 5 per cent of the resources provided through extension services in Africa are available to women, notes Ms. Marie Randriamamonjy, Director of the FAO's Women in Development Service, "Although, in some cases, particularly in food production, African women handled 80 per cent of the work. Of total extension agents at work in Africa today, only 17 per cent are women." There has been some growth in the number of non-governmental organizations and women's associations involving or working with rural women. Sometimes these are mixed organizations, but frequently, rural women prefer to belong to groups run by women. Traditional credit programs have failed to reach these farmers. Micro-credit as bottom-up rather than top-down financing has potential for sustainable and gender-neutral rural financial systems.

Co-operatives offer an alternative model of social organization

Co-operatives all over the world offer an alternative model of social organization which addresses some of the core problems within contemporary capitalism such as inadequate employment, global inequality and food insecurity (Wolff, 2012). In the past they offered another way of organizing businesses within a predominantly capitalist economy, but as the inherent weaknesses of capitalism are becoming more apparent, many regions in different parts of the world are adopting the successful approaches of Mondragon Corporation in Spain's Basque region. It is a corporation and federation of worker co-operatives. Mondragon Corporation is the largest business group in the Basque region and the seventh largest in Spain. Mondragon has a pay cap in which the highest paid worker can only earn 6.5 times the lowest paid; it has 85,000 members of which 43% are women. This one company reduces gender and income inequality in the Basque region. Mondragon shows how co-operatives growing to large sizes in many economic sectors can be transformative. Similarly *La Cooperativa Humar – Marinaleda*, a farmer co-operative in the Andalusian region of southern Spain, pays all of its workers the same € 47 per day and they earn almost twice Spain's minimum wage. Jobs are rotated and their motto is "To work less so that all may work." Contrast this to the situation in many developed countries when many high ranking people are proud to have several jobs, including paid board memberships, even at times when unemployment figures are very high. The Marinaleda council implemented measures to prevent the real estate speculation which undermined the rest of the Spanish economy. The co-operative has also provided education and has an action squad instead of a legislative body (Anon, 2012). The author of the article on Marinaleda calls it utopian and a genuine solution to market forces.

References

ACCORD (2003). *Canadian PM announces big $$$ for Social Economy*. Retrieved October 1, 2012, from http://www.accord.org.au/social/infobriefs/canadianpm.html.

Anon (2012, 29 August) Workers' cooperative defies crisis. *Presseurop* (English) Retrieved from www.presseurop.eu/en/.../2601741-workers-cooperative-defies-crisis.

Anon (n.d.). *Gender and Development*. Retrieved October 1, 2012, from http://go.worldbank.org/A74GIZVFW0.

Arthur, L., Scott-Cato, M., Keenoy, T. & Smith R. (2003). Developing an operational definition of the social economy. *Journal of Cooperative Studies*, 36, 163–189.

Chaves, R., & Monzón, J.L. (2000). Las cooperatives en las modernas economías de Mercado: perspectives españolas. *Rev Economistas*, 83, 113 – 123.

Chikwendu, D.O ., & Arokoyo, J.O (nd). *Landownership and access to farm inputs by rural women in Nigeria*. Retrieved October 1, 2012, http://www.fao.org/DOCREP/V9828T/v9828t08.htm.

Corcoran, H., & Wilson, D. (2010). *The worker co-operative movements in Italy, Mondragon and France: Context, success factors and lessons*. Retrieved from Canadian Social Economy Hub, Project No. Hub 20 website http://www.socialeconomyhub.ca/?q=content/occasional-papers

Côté, Daniel, Murray Fulton and Julie Gibbings. (2000) "Canadian Agricultural Cooperatives: Critical Success in the 21st Century – Summary Report." Retrieved from Canadian Cooperative Association website http://www.coopcca.com/agricoops/Summary%20Report1.pdf\

Côté, D., & Fournier D. (2005). Women & Social Economy. Is Quebec's 'Third Way' gender-sensitive? *Making Waves* 16 (3) 58 – 61.

w3.uqo.ca/oregand/publications/dc2005MakingWaves.pdf

Côté, D., Gagnon, E., Gilbert, C., Guberman, N., Saillant, F., Thivierge, N & Tremblay, M. (1998). *Who will be responsible for providing care? The impact of the shift to ambulatory care and of social economy policies on Quebec women.* Ottawa, Ontario: Status of Women Canada, 256 pages. View SW21-31-1998E.pdf

Cox, L., & Mullan, C. (2001). Social movements never died: Community politics and the social economy in the Irish Republic. In: International Sociological Association / British Sociological Association *Social Movements Conference*, November 2001, Manchester. http://eprints.nuim.ie/1529/

FAO (2001). Commodity Market Review, 1999-2000, Commodities and Trade Division, Food and Agriculture Organisation of the United Nations. Rome. Retrieved from FAO website http://www.fao.org/docrep/003/X7470E/X7470E00.HTM

Favreau, Louis (2008). *Entreprises collectives, les enjeux sociopolitiques et territoriaux de la coopération et de l'économie sociale*, PUQ, Collection Pratiques et politiques sociales et économiques. http://www.dec-ced.gc.ca/eng/publications/agency/evaluation/179/page-4.html

FedNor (2005). Ontario Social Economy Consultation. Submitted to FedNor/Industry Canada by Tamarack – An Institute for Community Engagement, Feb 2005.

Fontan Jean-Marc, Shragge, E. (2000). Social Economy: International Debates and Perspectives. Black Rose Books. http://www.web.net/blackrosebooks/social3.htm

Geller, S., & Salamon, L. (2007). Non-profit advocacy: What do we know? John Hopkins Center for Civil Society Studies. CCSS Working Paper CCSS-WP-22. Retrieved from http://www.advocacyinitiative.ie/download/pdf/ccss_wp22_2007.pdf

Hansmann, H. (1999). Cooperative Firms in Theory and Practice. *Finnish Journal of Business Economics,* 4, 387-403.

Lans, C. (2005). What is the Social Economy? Victoria: British Columbia Institute for Co-operative Studies. Formerly available at http://www.socialeconomyhub.ca/hub/index.php?page_id=9.

Laville, J., Levesque, B., & Mendell, M. (2008) Diverse Approaches and Practices in Europe and Canada in *The Social Economy: Building Inclusive Economies*, OECD Publishing. doi: 10.1787/9789264039889-7-en

Lewis, M. (2004). CED and the social economy break through onto the federal agenda. Making waves, Spring 2004. http://www.cedworks.com/endofthebeginning.html

Miller, E. (2006). Other economies are possible. *Dollars & Sense: The Magazine of Economic Justice*. Reprinted on September 2, 2006 on Alternet http://www.alternet.org/story/40339/

Monzón Campos José Luis (1997). Contributions of the social economy to the general interest. *Annals of Public and Cooperative Economics,* 68 (3), 397 – 408.

Moran, W., Blunden, G., & Bradly, A. (1996). Empowering family farms through cooperatives and producer marketing boards. *Economic Geography,* 72, 161 - 177.

Moulaert F, Ailenei O. (2005). Social economy, third sector and solidarity relations: a conceptual synthesis from history to present. *Urban Studies* 42(11), 2037-53.

Neamtan, N., & Rupert Downing, R. (2005). Social Economy and Community Economic Development in Canada: Next steps for public policy. Montreal: Chantier de l'économie sociale, in collaboration with CCEDNet and ARUC-ÉS. http://politiquessociales.net/IMG/pdf/issues_paper21sept_final.pdf

Neamtan, N. (2002). The Social and Solidarity Economy: Towards an 'Alternative' Globalisation. (A. Mendell, Trans.). Background paper – in preparation for the symposium, *Citizenship and Globalization: Exploring Participation and Democracy in a Global Context.* Sponsored by: The Carold Institute for the Advancement of Citizenship in Social Change, Langara College, Vancouver, June 14-16, 2002. (pdf)

Perkins, Daniel; Nelms, Lucy and Smyth, Paul. Beyond Neo-liberalism: The Social Investment State? [online]. Just Policy: A Journal of Australian Social Policy, No. 38, Dec 2005: 34-41. Retrieved from: http://search.informit.com.au/documentSummary;dn=178615069694270;res=IELHSS> ISSN: 1323-2266.

Prentice, S., & McCracken, M. (2004). *Time for action: An economic and social analysis of childcare in Winnipeg.* Winnipeg, MB: Child Care Coalition of Manitoba.

Quarter, J., Mook, L., & Richmond, B.J. (2003). What is the Social Economy? *CUCS Research Bulletins* Number 13, March 2003.

Rodrik, D. (2002). Trade Policy Reform as Institutional Reform. In B. M. Hoekman , P. English, and A. Mattoo (Eds.), *Development, Trade, and the WTO: A Handbook,* World Bank, Washington, DC, 2002. Retrieved from http://www.hks.harvard.edu/fs/drodrik/Research%20papers/Reform.PDF

Rodrik. D. (2001). Development Strategies for the 21st Century, in Annual World Bank

Conference on Development Economics 2000, Fall 2001. Available from http://www.hks.harvard.edu/fs/drodrik/Research%20papers/devstrat.PDF

Salkie, F. (2005). Defining the Social Economy. Access West, Apr-June 2005 (pdf 564KB)

Sätre Åhlander, A-M. (2000). Women and the social economy in transitional Russia. *Annals of Public and Cooperative Economics*, 71 (3), 441 – 465.

Sätre Åhlander, A-M. (2001). The social economy: new co-operatives and public sector. *Annals of Public and Cooperative Economics*, 72 (3), 413 – 433.

Smith, J. & McKitrick, A. (2010). Current conceptualizations of the social economy in the Canadian context. *Canadian Social Economy Hub* at the University of Victoria. Retrieved from: http://www.socialeconomyhub.ca/sites/default/files/DefinitionPaper.pdf

Government of Canada (2004), Speech from the Throne, February 2004. Speech from the Throne to Open the Third Session of the 37th Parliament of Canada. Retrieved from http://www.pco-bcp.gc.ca/index.asp?lang=eng&page=information&sub=publications&doc=aarchives/sft-ddt/2004_1-eng.htm

Government of Canada, (2004a), Speech from the Throne, 5 October 2004. Speech from the Throne to Open the Third Session of the 38th Parliament of Canada. Retrieved from: http://www2.parl.gc.ca/parlinfo/Documents/ThroneSpeech/38-1-e.html

Toupin, L. & Goudreault, N. (2001). The changing role of the state, women's paid and unpaid work, and women's vulnerability to poverty. Social and Community indicators for evaluating women's work in communities. Status of Women Canada, Ottawa, Ontario.

Westlund, H., & Westerdahl, S. (1997). Contribution of the social economy to local employment. Swedish Institute for Social Economy (SISE), Östersund.

Wilkinson, P. & Quarter, J. (1995). A theoretical framework for community-based development. *Economic and Industrial Democracy*, 16 (4), 525 - 551.

Wolff, R. (2012, 24 June). Opinion: Yes, there is an alternative to capitalism: Mondragon shows the way. Retrieved from http://www.guardian.co.uk/commentisfree/2012/jun/24/alternative-capitalism-mondragon

Voluntary Sector Initiative (Canada). Wyatt, B., & Kidd, M. (2003). *Strengthening Canada's charitable sector: Regulatory reform*. Ottawa, Ont.: Voluntary Sector Initiative, Joint Regulatory Table.

On how the media influence Canadian democracy

In this essay I illustrate how the consolidation of the media has had a negative influence on Canadian democracy. I define democracy as the ability of all, even the little guy, to have her viewpoint represented in the national media and to have the government look after her interests, and not only the interests of the multinational companies. I am not calling for "advocacy journalism" (a right wingers epithet for a progressive viewpoint); nor do I want to call for "fair and balanced" reporting now that Fox News has hijacked that famous phrase. Perhaps "equal opportunity, equal access" is the best motto for a newspaper that aims to promote democracy.

First some background. I left Canada some months before Marc Lepine shot 14 women in the engineering wing of L'Ecole Polytechnique in Montreal. I had finished my BSc at the University of Guelph and worked for only part of the year allotted to foreign students to work in Canada. I came back in 2001 with my PhD, a little sceptical about Canada being open to (willing to recognise and reward the credentials of) non-British immigrants. In the beginning I lived with relatives in that part of Ontario constantly portrayed in the Toronto Star as a hotbed of immigrant violence (Brampton) and I armed myself with the Globe and Mail, the Toronto Star and the National Post to aid my job search.

It was not long into the muggy Brampton summer before I realized that the National Post was no leftist newspaper and I stopped reading it. I even refused free copies on the UVic campus or with my coffee from the gas station (I left Ontario after 6 months, buying a one-way ticket to the other side of Canada to get away from Mike Harris and his devil-take-the-hindmost-revolution). Columnist Stephen Gowans wrote that Dalton Camp called the National Post "a 250,000-word editorial wrapped around a newspaper." However articles on organic agriculture posted to the organic listserve I subscribe to have come more often from the Post than from the Toronto Star, even if most of the Post's articles refer to the economic potential of organic farming due to the higher prices paid to the farmers. The business slant of the stories reflects the observation made by Dalton Camp in a lecture at the University of Waterloo on March 23,

2000: "today, half the front page of your daily paper is about money." His comment applies not only to Lord Almost's paper but could also refer to the Toronto Star of March 20; "Bird flu could shave $60B off GDP, manufacturers say" was the headline of one Canadian story that never mentioned the word farmer.

The Post belongs to the late Izzy Asper's megalithic CanWest Global Communications along with newspapers such as the Victoria Times-Colonist which often carries the same articles as the Vancouver Sun [one downside of media consolidation]. The media giant also owns several smaller publications but not Vancouver's independent Georgia Straight (circulation 120,000). The Straight printed an investigative report (December 2004) on a former public servant who lost her job for requiring farmers, including relatives of the then BC Minister of Agriculture, to adhere to the environmental regulations of the province. By the way that happened to be her job. I reprinted most of the story in the Sierra Club of Canada Victoria newsletter because I did not see the story in any of the CanWest newspapers. This does not mean that I consider myself one of the "few good women" that Dalton Camp claimed were "protecting the public virtue and its interests" in his May 2001 commentary (Southam Press); although the public servant Bev Anderson definitely was. I decided to use my temporary economic independence as a post-doc to counter the news suppression effects of media consolidation.

A story about the Georgia Straight ran in the back pages of the provincial media on the same day that media baron Izzy Asper was being eulogized across the media landscape. The story was that the BC Liberal government had decided that this and only this alternative weekly was not a "newspaper" because it exceeded the 75% limit in space devoted to ads. They were going to remove its tax-exempt status and impose a $1 million back-tax penalty and were only stopped by a public outcry.

The BC Liberal government, in the opinion of many whose views are seldom reflected in newspapers (social activists, feminists, the unwaged) are only masquerading under that name and are really neo-cons. I doubt that they attended or would have agreed with the March 2000 lecture given by Camp as Stanley Knowles Visiting Professor of Canadian Studies called "Neo-Conservatism: How to Wreck a Country with a Hammer, Part II." Judy Rebick quipped on Rabble News (March 2002) that Camp received his transplanted heart from a socialist donor, which is why he wrote his political commentaries as a Red Tory while his contemporaries moved to the Thatcher/Reagan/Bush-led right.

The Toronto Star does not provide equal coverage to small farmers or organic farmers. The last major organic story covered was in 2002 when canola farmers in Saskatchewan had the "cojones" to sue Monsanto over genetic contamination. The unbalanced coverage is very evident in the current hot topic of avian flu. Some farmers have resorted to websites and blogs to tell their stories but those outlets do not have the same impact as a national newspaper (unless of course you are Arianna Huffington compiling a fake blog by George Clooney).

The Globe and Mail newspaper (which like The Star, I read on-line every day) allows 90 day searches and below are some recent headlines on avian flu revealed by the search:

1. Five countries report presence of avian flu virus
2. Don't wait till it's too late: CME. Plan now for a bird flu pandemic, report urges, it could devastate your work force
3. Avian flu killed five youths in Azerbaijan, WHO says
4. Have experts made too big of a squawk about bird flu?
5. Bird flu picks bad place to spread

Only the last two articles, after a year or more of pandemic headlines, start raising questions on the level of the threat posed by the virus.

Internet searching for newspaper articles in the Canada.com search engine belonging to CanWest produces more hits - 500 articles from across Canada about bird flu. One of these from the Ottawa Citizen has the most interesting headline: "Pandemic paranoia: Rumsfeld's connection to Tamiflu has conspiracy theorists spinning"

> "One day Americans were panic-buying Tamiflu, making the influenza remedy the most sought-after drug in the world; the next day, rumours were rampant that the U.S. response to bird-flu was more about profit than public health."

This conglomerate-owned newspaper article acknowledged what was reported in the March 12 London Independent story by Geoffrey Lean and Jonathan Owen entitled "Donald Rumsfeld makes $5m killing on bird flu drug." The story opens thusly:

"The US Defence Secretary has made more than $5m (£2.9m) in capital gains from selling shares in the biotechnology firm that discovered and developed Tamiflu, the drug being bought in massive amounts by Governments to treat a possible human pandemic of the disease."

Even CNN Money (October 31, 2005), no leftist newspaper, reported on the links between Rumsfeld and Tamiflu. Including in their story that:

"former Secretary of State George Shultz, who is on Gilead's board, has sold more than $7 million worth of Gilead since the beginning of 2005"...............................

"What's more, the federal government is emerging as one of the world's biggest customers for Tamiflu. In July, the Pentagon ordered $58 million worth of the treatment for U.S. troops around the world, and Congress is considering a multi-billion dollar purchase."

According to the Toronto Star avian flu is a problem linked to wild birds and "backyard flocks kept by families" that are the "Province's Achilles heel" (March 1). Several newspapers neglect to make the distinction, made with all credit to the Toronto Star (March 12), that bird flu has killed or forced the slaughter of millions of chickens and ducks. Several writers only write "killed" or "decimating" (March 11) giving a completely false impression.

The Star covered large Ontario poultry farms and their biosecurity protocols (March 20). One other Canadian-focussed story (March 2) was about the quarantined ducks in Quebec which tested negative for the H5N1 virus. Not a word was quoted from those owning the ducks. Nor was there any coverage of the Quebec farmers who now have to keep their flocks indoors and thus risk losing their organic certification and their livelihoods. Quebec shares a provincial border with Ontario and its citizens should be considered newsworthy by The Star. The Star found the following more interesting: Austrian cats with bird flu (March 6), data hoarding by labs (March 12), Roche drug capacity (March 16). This leads me to concur with columnist Molly Ivins who wrote on AlterNet on March 23, "for some reason, publishers assume people will want to buy more newspapers if they have less news in them and are less useful to people."

Meanwhile alternative viewpoints like those of wildlife biologist David Hancock are carried in independent outlets like Vancouver's Bannerline Communications. Hancock spoke to Dr. George Luterbach, the Director of Veterinary Services CFIA Western Division who confirmed that wild birds were the vectors of the H7N3 found in BC in 2004 but that wild birds did not carry the lethal form of the virus. Luterbach said the issue boiled down to getting the commercial industry exporting chicken products as soon as possible and that the CFIA had to eliminate the perception that they had not done everything possible. Hancock however linked the spread of avian flu in BC to the dumping of poultry litter including carcasses in fields by commercial farmers, which would be picked up by wild birds when feeding.

The venerable BBC reports a similar theory on the spread of avian flu in a report by Dr Leon Bennun (February 17):
"countries like Japan and South Korea, which imposed strict controls on the import and movement of domestic poultry after initial outbreaks, have suffered no further infections............countries which have not yet developed a large-scale intensive poultry industry have also been largely spared. The FAO reports that in Laos, 42 out of 45 outbreaks affected intensive poultry units."

Doesn't The Star read its own headlines? On March 20 The Star quoted a Belwood poultry farmer who said "bird flu's always been around." Given the recent multi-part headline in the Toronto Star (March 23, 2006), "Scientists find bird flu clue - Why virus limited in spread to people - Not easily coughed or sneezed out"; is it not in the public interest to investigate if the Canadian health service is investing taxpayer dollars in a pandemic that may not materialise but has only become important due to extensive media coverage that [who knows] may have been shaped by the extensive political connections of the biotech company with a patent on Tamiflu? Even conspiracy theories need investigation [and could sell newspapers].

An investigative report on avian flu was done by GRAIN, an international non-governmental organisation that promotes the sustainable management and use of agricultural biodiversity. In a post about the GRAIN report to the organic listserve I mentioned earlier, an organic farmer mused on the non-appearance of this report or even the sustainable agriculture viewpoint in the major newspapers.

A musing that brings me full circle - the consolidation of the media has had a negative influence on Canadian democracy. Given the definition of democracy as the ability of all, even the little organic guy, or wildlife biologist, to have his/her viewpoint represented in the most read newspaper in Canada. I echo Camp's call for more "meliorists" which doesn't mean advocacy journalism, but "optimists who believe something can be done about improving things." An activist who embodies my interpretation of that oft-quoted Edward Bulwer-Lytton expression, "the pen is mightier than the sword [of neoliberalism]."

Figure 1. Dutch goat showing animal welfare

Methods in EVM - North America and the Netherlands

Dr. Cheryl Lans and Drs A.G.M. (Tedje) van Asseldonk*

*IEZ: Institute for Ethnobotany and Zoopharmacognosy

Abstract

Research in Veterinary Anthropology in Europe and North America is best conducted using a participatory data collection and dissemination method. This method bridges the gap between scientists and knowledge holders (farmers), and it also generates a large amount of usable data. The method is ethically-sound and gender-friendly and it allows the participants (farmers, herbalists, alternative practitioners) to improve themselves economically and socially through mutual self-help. It is a collaborative venture in that participants can pool resources, abilities and information thus multiplying the likelihood that they can obtain useful solutions and it minimizes the risk of failure.

Keywords: ethnoveterinary medicine, dissemination, participatory workshop method, American colonial medicine, Native American medicines, The Netherlands

Introduction: Debates about applied anthropology

Ethnoveterinary medicine is the scientific term for traditional animal health care, the study of which has also been called Veterinary Anthropology. Ethnoveterinary research is typically conducted as part of a community-based approach that serves to improve animal health and provide basic veterinary services in rural areas. Veterinary anthropology studies folk management of animal health in the context of the whole farming system, taking into account other socio-economic and political considerations such as cultural beliefs, religion, societal norms and trends and by the *materia medica* available in terms of local flora, fauna and minerals [1; 2]. Ethnoveterinary practices are usually tested to assess their validity, but the majority of the studies are descriptive. It has been a named and recognized area of academic interest since the mid-1970s.

Ethnoveterinary practice in the Netherlands is nearly completely restricted to folk medicine [3], as there were no registered Dutch animal natural health practitioners until recently. Since 1998 two private schools for the natural health care of animals have been started. A professional organization of graduates has come into existence since 2003 and is called VBCND (website: www.bcnd.nl).

Animal scientists and farmers are forced by society to be practical. They have to produce food animals in an affordable way without affecting the animal's growth or the health of the future consumers of the meat or other products (eggs, dairy products). They also need to take animal welfare, environmental contamination and food residue concerns into account (Figure 1). Risk assessment of veterinary medicinal products in relation to residues involves the application of more stringent criteria than in any other biomedical discipline [4]. It is due to these ongoing factors that ethnoveterinary medicine offers fitting solutions for animal-based food production.

Medicinal plants are used by small-scale organic livestock farmers who are restricted from using allopathic drugs or by farmers whose economic circumstances prevent the use of veterinary services for minor health problems of livestock. For example the average farm income in the Netherlands in 2009 decreased to 15,100 euros per farm, this is nearly the lowest acceptable income for Dutch citizens (€ 1284,-/month in 2009); its lowest point in twenty years [5]. As a result of earning at the social minimum for living in their country, farmer holdings undertake more and more extra activities, take jobs outside the farm or give up. So they need cost-effective anthelmintics and other alternative plant-based medicines in order to provide the organic meat that consumers increasingly want. Even animal health professionals have recognized that veterinary expenditures represent a large part of animal production costs, and therefore their reduction or containment is essential for the maintenance or improvement of a farmer's income [6]. A summary of the Dutch history of enlargement and empoverment of farms can be found on www.dutchexperience.com.

Ethnoveterinary research confirms centuries of medicinal plant use by people across Europe so it cannot be said to be at variance from accepted knowledge and thus needing extraordinary proof. Germany has the highest rate of acceptance of herbal knowledge in the Western hemisphere.

Ethnoveterinary medicines used in Canada are mainly derived from Europe, First Nations and American Indian traditions and from Asia. One of the purchased products used for cancer in pets in British Columbia, Canada is based on a formula popularized by Canadian nurse Rene Cassie. This "Essiac Formula" is reported to be: 6 1/2 cups burdock root – cut (*Arctium lappa*), 16 oz. sheep sorrel herb – powdered (*Rumex acetosella*), 1 oz. turkey rhubarb root – powdered (*Rheum palmatum*), 4 oz. Slippery elm bark – powdered (*Ulmus fulva*). The 8 herb formula was given to Mrs. Johnson by an Ontario Ojibwa medicine man. Mrs. Johnson then told the remedy to Rene Cassie (Figure 2) [7; 8].

Figure 2. Nurse Rene Caisse, promoter of the Essiac herbal formula. Source: bracebridge.library.on.ca

Applied anthropology

Ethnoveterinary research in developed countries attempts to create the necessary space for the folk medicinal paradigm to co-exist with western medicine. A framework of indigenous / ethnoveterinary knowledge that can interface with science and technology is more likely to influence scientific research agendas and development work [9]. This type of research fits into the framework of applied anthropology, which is a system of research-based, instrumental methods which introduce change or stability in specific cultural systems through the provision of data, initiation of direct action and/or the formation of policy [10]. Applied anthropology has been described as a process rather than a field [11]. In this process the relevant stages are discovery/planning, intervention and evaluation.

Applied anthropology has also been defined as analyses of particular human problems, situations, or processes for the purposes of understanding their causes, dynamics, and consequences. Research results can then be used to positively influence the research area [12]. Harnessing anthropology to technical knowledge in order to facilitate development puts the discipline at the centre of the development process before others "steal Anthropology's disciplinary clothes" [13].

A few decades ago anthropologists selected research projects that would do little harm or provide some minimal benefits [14]. These researchers tried to exercise as little influence as possible on the social phenomena they studied and left the active intervention in the situation to other agencies [14]. Positivistic approaches in anthropology are based on three deeply embedded views of its objectives:

1. Knowledge accumulation is an end in itself,

2. Studies are preferably done on tribal and peasant societies to which anthropologists have easy access. These more or less autonomous, traditional communities do not change while contemporary society, which does change, is not as suited to anthropological study. Unchanging societies provide the theoretical basis for the evolutionist, diffusionist or functionalist anthropologists who assume the existence of more or less closed and finite social systems in order to demonstrate their theories of social equilibrium [14],

3. Long-term studies are preferable to short-term analyses of cultural systems [14; 15]. These long-term studies are seen by anthropologists as one of the strengths of anthropology but they have also been termed one of the weaknesses that feeds into the cultural relativism and political correctness [16]. Academics may not want to have their monographs challenged by development practitioners, and may not want to have to revise them too often as a result of cultural change. "Thick descriptions" have often been challenged by native readers [17]. They have been called 'inscriptions'...which 'can be recombined in various ways, communicated

across space and time and they thus become very important and significant', more so than the original knowledge holders [18]. The critiques from native anthropologists have led some Western anthropologists to become self-referential [19].

Some anthropologists did not agree that researchers should use his/ her knowledge for a particular cause, or try to include ethnobotany into science [20; 21; 22; 24], while others wrote about local knowledge as social resilience [23]. Some indigenous knowledge researchers wrote that if indigenous knowledge systems represent the cultural dimension of development, then they cannot be reduced to the empirical knowledge that they contain. If a culture is a system as other anthropologists claim, then when a cultural item is removed this removal disrupts the set of relationships into which it is locked, and the cultural item cannot be imported into another culture without bringing with it some of the trappings of those former relationships and disturbing its new surroundings, therefore cultural change is problematic [25]. These statements ignore the fact that communication across cultural boundaries, and plant dispersal has been happening for millennia, but culture/system anthropologists ignore this truism [25].

For example many Native American ethnobotanical remedies were transferred to Europeans in North America; for example the selling of *Echinacea* spp., patent medicine for snakebites is said to be the origin of the term snake-oil salesman [26]. Early colonists adopted the use of Oregon grape tonic [27]. American native medicinal uses of *Lobelia* species and the use of foxglove as a cardiac stimulant pre-date European arrival [28]. Native Americans were reluctant to share their knowledge (for example the use of *Rumex verticillatus* to treat ulcers) except with trusted missionaries and allies. He credits United Brethren missionaries like John Heckewelder for eighteenth-century accounts of the Delewares medicines (barks of white walnut, cherry, maple and birch) [28]. Settlers were making studies and sending reports back to Europe. For example *Artemisia dracunculus* was considered to be a native American plant by Dutch settlers in New York in 1650, while others do not agree with this interpretation of the Dutch report [28].

Some practitioners face limits when taking a critical approach in anthropology. Some clinical anthropologists are asked to function solely as cultural interpreters and when they "study up" their critiques of health care arrangements and their political-economic environments are not welcomed [29; 30].

In the Netherlands, Wageningen University and Research (WUR), the former agricultural university, has played a leading role in applied agricultural research. WUR has for the past two decades used an approach called actor-oriented sociology. It essentially focuses on the analysis of strategic behavior and on why people do not work together. Röling [31] has pointed out that this approach has detracted from the social contract of Wageningen social sciences because it focuses on the reasons why people make selfish choices in social dilemmas while neglecting the conditions under which people make co-operative choices, and it also dissuades graduate students from undertaking participatory development, collective action and other projects that would lead to a sustainable environmentally-sound future. Society is the victim of the distancing of science from social change since research outcomes are often left forgotten on some dusty shelf or used only to call for further theoretical research [32].

Although WUR research chairs are increasingly working together there is no indication that they have embraced the particular participatory method that we want to highlight in this review. Post-structuralists, including those based at WUR, are acting as constraints to development anthropology because they are unwilling and incapable of acting politically "because they stand for a rather rigid form of cultural relativism and political correctness" [16]. They dismiss agents of change from the West but cannot suggest alternatives because their theoretical approach does not allow them to intervene; they can deconstruct but not more. However their hands-off approach is a farce because the community-centred praxis approach recognizes that all action, even the casual observation characteristic of positivistic research affects a system and that inaction is consequential and thus ultimately partisan [11; 33]. Action/praxis research takes this insight within its approach [32]. Each researcher is an actor in the research process.

Rappaport has called for an engaged anthropology that would correct problems [12]. Anthropology's traditional tools—such as ethnographic interviews and participant observation— are thus employed within the framework of a more general research action methodology [34]. Ethnography thus provides the basis for intervention and planned change, advocacy and information dissemination as well as theory development [35]. Cox [36] warns applied researchers that they may encounter conflict with community power groups and brokers in the field and also a loss of professional peer support in academia [37; 15]. Applied work is often challenged because there is a split between theory and practice which is said to be linked to masculine ideas of science which exclude advocacy and activism from "real science" and determine what counts as publishable material [38; 39].

Engaged research in medicine

Baer sets out four premises for a Critical Medical Anthropology [29; 30]:
1. The recognition that local events can be influenced by external forces.
2. A holistic understanding of sickness and an examination of the power relationships that exist in medicine.
3. The realization that all theories have a cultural basis, including the anthropological theories of sickness and healing.
4. The acceptance of research and theorizing as social acts that can be made into ethical acts.

The [US] health system would be better served by physician engagement in a medical commons, ideally with communities of consumers, and this is arguably the only approach that will ensure proper allocation of health care resources [40]. Lawrence [41] published a letter in the journal Current Biology claiming that the current competitive structure of modern science is more suited to those who are prepared to show off and to exploit others while modest and less aggressive men and women are not given opportunities to flourish even though they are just as capable of conducting rigorous science. He claims that even though there are more female university students than in previous decades, they are pushed out due to the unnecessarily competitive nature of science and that science loses many original researchers as a result. He thinks that science would flourish in a more understanding and empathetic workplace.

Clinicians believe that clinical practice must be evidence based and that randomized controlled trials and secondarily systematic reviews provide the best evidence upon which to base medical practice [42]. In this world-view however, observational studies and accounts of accumulated clinical experience are not considered meritorious or interesting, and so, according to the conventional wisdom, they are not published. Observational studies have merit especially for personalized medicine and better standards for them should be designed [42]. Anecdotes are frequently published in compilations of the side effects of drugs and provide information of the potential adverse reactions or interactions, mechanisms, diagnostic techniques, or methods of management; anecdotes can generate or test hypotheses, and remind or educate; and, like trials, they can be subjected to a kind of systematic review [43].

The first North American medical journals contain many anecdotes, observational studies and early trials of North American native plants. President Jefferson was a pragmatist interested in the practical benefits of nature not in obscure scientific theories that served a narrow audience. This led him to fund the expedition of Meriwether Lewis and William Clark [44]. President Jefferson suggested that Meriwether Lewis take note of the remedies of the Native Americans along the Missouri river; *Hydrastis canadensis* was one of the plants that Lewis noted. Ethnoveterinary research undertaken in British Columbia, Canada revealed that *Hydrastis canadensis* had many uses for pets (abscesses, bites, wounds, skin problems, and infections) that reflected the Native American antimicrobial use [45]. Scientific studies of the plant beginning in 1798 may have pushed it to prominence [26; 28]. Alfred Durand noted in the *The American Journal of Pharmacy* the native use of goldenseal and claimed that a tincture was being used in North America as a tonic, while the infusion was used topically in ophthalmic and ulcerous inflammations [46]. Robert Sattler M.D. wrote that preparations of *Hydrastis canadensis* were well known "as therapeutic agents for the various affectations of mucous surfaces" ... he tested "its efficacy in the management of catarrhal processes of the nose, retropharyngeal space, and pharynx" but found that however useful it might be, it produced a yellow stain [47]. An editorial in the American Medical Journal claimed that *Hydrastis canadensis* enjoyed an almost enviable reputation among the eclectic practitioners, homeopaths and liberal practitioners of all schools [48].

These medical studies make nonsense of the anthropological claims about indigenous knowledge systems which cannot be reduced to the empirical knowledge that they contain; and the claims that cultural items cannot be removed from the set of relationships into which they are locked, and imported into other cultures [20 - 25]. Scholars are also restricted if they do not use myths and folk tales. For example *Achillea millefolium* was thought to be native by John Josselyn who visited New England in 1638 and by John Brickell who published a book on the natural history of North Carolina in 1737 [28]. *Achillea millefolium* was chiefly used as a wound plant in Europe, deriving its name from the Greek hero Achilles who reportedly staunched the wounds of his soldiers with it.

Figure 3. Participants at an ethnoveterinary workshop at the University of Victoria in 2006

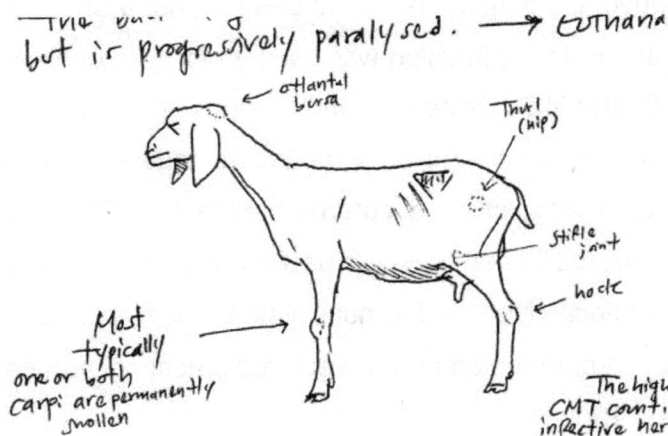

Figure 4. Illustration done by one of the participants at an ethnoveterinary workshop at the University of Victoria in 2006

Should research participants be rewarded?

Contributions to the entire community such as the Raika Fund set up by camel ethnoveterinary specialist Ilse Kohler-Rollefson [49] have been cited as appropriate by Srivastava [50], since the specialists involved have to spend considerable time explaining their techniques to the anthropologist. The specialists also took time to learn and develop these indigenous practices. Two problems arise from the non-participatory and experimental-science-driven approach that university science practices in the field of ethnoveterinary medicine.

- If farmers are to benefit from the work of agricultural scientists they need to be better involved in the processes of documenting and evaluating their traditional knowledge.

- Ethical questions arise: what is done with farmer's knowledge that stands the test and is of value outside the ethnoveterinary applications? Modern biopiracy can result; as can be illustrated by the fact that herbs like oregano and thyme (*Origanum vulgare* L. /*Thymus vulgaris* L.) that have more than 2000 years of history of use in Europe are also claimed by several patent holders for uses of herbs or the main oil components carvacrol and thymol. Recently scientists of Veterinary Faculty of Utrecht University in the Netherlands had a government funded project that was used to register patents on plant compounds to be used as antibiotics.

See the following websites for more information:

www.dutchfarmexperience.nl

The Fyto-V group website is www.fyto-v.nl ; the final report (ref 60) can be found on the WUR site

http://www.wageningenur.nl/nl/Publicatie-details.htm?publicationId=publication-way-333737353634

The stable books in English version are only on the website www.fyto-v.nl but to find them you have to look not in the english part but in the dutch part, where both dutch and english versions are ; they can be found in

http://www.fyto-v.nl/docs/sb_dairy.pdf

http://www.fyto-v.nl/docs/sb_pigs.pdf

http://www.fyto-v.nl/docs/sb_poultry.pdf

They unsuccessfully sought commercial partners to exploit these patented applications exclusively, while both in the Netherlands and in other European countries farmers have been using oregano oil and other aromatic plant products for this purpose for many years.

One of the reasons that scientists publish quantitative studies and apply for patents is that they need to choose their research projects based on career rewards – funding, papers and academic positions (the "context of pursuit") [51]. Bauer [52] confirms the points that Janssen [51] makes about the social nature of scientific advances (with the use of terms like resistance to discovery, premature discovery and delayed recognition). Science, he writes, advances on three fronts: the observation of striking new phenomena, the introduction of new methods, and the development of new theories; deviating from normal science in two of the troika marks a "high-risk" venture that most scientists avoid. "*Drastic change in any aspect is resisted; proposed simultaneous change in two of them causes a discovery to be neglected, isolated from the mainstream action. Claims that change is needed simultaneously in all three aspects tend to be dismissed as pseudoscience*".

The progress of science can be deliberately guided or accelerated [52]. This is because science is an intellectual activity (internalist factors) influenced by the aspirations and emotions of those who practice it and by the institutions that they have developed, which in turn; interact with other social institutions (externalist-social factors)... "What makes discoveries really significant is "demonstrating them in a way that convinces the scientific and technical establishment" ... "The important part of a scientific discovery in almost any aspect of science is the reception it receives."

Bauer [52] continues his paper with a discussion of the different social factors that affect the progress of scientific research. His first category is *Resistance and neglect. Resistance to scientific discovery may be passive or active: passive, if a claim is ignored, not investigated, set aside; active, if the claim is opposed or pronounced mistaken.* It is not resistance *per se* that should be seen as hindrance to the acceptance of fields and methods such as ethnoveterinary research and participatory methods but only *excessive, unwarranted* resistance [52].

For example, in 2011, $210 million euros were given in aid to vegetable farmers in Europe affected by lost sales due to enterohemorrhagic *E. coli* (EHEC) which only compensated them for 50 - 70% of their losses due to unsold vegetables (see the website http://www.spiegel.de/international/europe/0,1518,768534,00.html).

There are ethnoveterinary remedies involving plants that have efficacy against *E. coli*: *Usnea longissima, Urtica dioica, Althaea officinalis* and *Cinnamomum zeylanicum*. The plants were being used to treat pets on farms [53] (farm pets are a potential reservoir of pathogens that affect humans [54]).

Another factor in support of ethnoveterinary medicines is that oil reserves (the source material for 20th century science and chemicals such as fenbendazole, albendazole etc.) have been depleted and future products derived from petrochemicals may be less available and ethnoveterinary alternatives could become valued replacements. One writer [55] notes that the world is now very close to peak oil and gas production, if we have not already passed it. It took the United States forty years to go from peak discovery to peak production, and twenty seven years for that movement in the North Sea. It is estimated that 60 Gb (billion barrels) of deep sea oil will only provide three years of world supply. Even though there are still discoveries the easily obtained supplies have been depleted [55].

Of great urgency at this moment is the problem of the increasing antibiotic-resistance in bacteria. Antibiotic resistance in the US (MRSA – methicillin resistant *Staphylococcus aureus*) kills more people than HIV [56].

"The antibiotics sold in this country for food animal use far outstrips the amount of antibiotics sold for human use," said Gail Hansen, a program officer with the Pew Foundation Campaign on Human Health and Industrial Farming. "About 30 million pounds (13.6 million kilograms) of antibiotics are sold for animal use; about 7 million pounds are sold for human use." [56].

Multi-resistance of bacteria due to the dissemination of the ESBL enzymes is moving forward rapidly. It is feared that the Netherlands, with a dense population of both people and livestock (the first sober, the latter abundantly treated with antibiotics), will suffer in the near future many bacterial and fungal infections that cannot be overcome by regular medicine. Currently 80% of the poultry meat is infected; in the human population with gastrointestinal minor complaints it is 10% but figures are rising, especially in places with high use of antibiotics [57 – 59]. So alternative ways of health-promoting and disease management are needed in this society, but this topic has not been taken up by any Dutch university. On the contrary there is no pharmacognosy or pharmaceutical botany in Dutch universities and Ethnobotany in anthropology is mainly performed in South-East Asia and Suriname.

The Institute for Ethnobotany and Zoopharmacognosy (IEZ)

IEZ is a private knowledge center committed to research, learning, and teaching about ethnoveterinary and herbal medicine. IEZ in collaboration with other researchers in ethnoveterinary medicine and conventional veterinary medicine provides community-based technological solutions to farmers' animal health problems based on traditional knowledge.

The Institute for Ethnobotany and Zoopharmacognosy has in the past already undertaken ethnoveterinary research. For example as a part of their training in herbalism 85 students of the two private schools for natural animal care that exist in the Netherlands were assigned to ask farmers, pet owners and animal pension keepers for their traditional remedies. This led to 168 case reports in the period 1998-2004, mainly regarding horses and small pets (See Figure 5). In the period 2006-2008 IEZ participated in the Fyto-V project that experimentally evaluated herbal remedies that are in the market (mainly as food ingredients or additives) for pigs, poultry and dairy farms. This government funded project was a request of the organic farmers association, that until that moment only had obtained scientific support from WUR Livestock Research (WUR-LR) in this respect, and wanted to involve specialists with more expertise on traditional herb use. The Fyto-V project was undertaken by Rikilt (the WUR-food safety institute), IEZ and other parties. The Fyto-V results were disseminated in 2009-2010 with the production of stable-books that advised on integrating the use of herbal and other natural products in farm management [60]. These booklets had a researcher-supplied approach.

Farmers in the Netherlands, in particular organic dairy farmers, possess traditional knowledge about herbs that can benefit the health of their animals. In 2011-2012 IEZ was involved in a cooperative project with Louis Bolk Institute, Wageningen University and HAS den Bosch that explored the knowledge several organic dairy farmers have of the biodiversity in their pastures and the way they value it to influence the health of their cows. Still the potentially innovative knowledge of Dutch farmers and other citizens on herbal medicines remains largely unexplored, due to the lack of proper methodology to do this in both a scientific and an ethically sound way and a lack of adequate expertise in this field in the Netherlands. Two projects planned by LTO Noord (a large Dutch farmers association) together with IEZ, aiming to enlarge the medicinal value of pastures with herbs, were submitted in 2012, but they were not funded by the Dutch government because the government claimed that the potential usefulness of medicinal herbs was unproven.

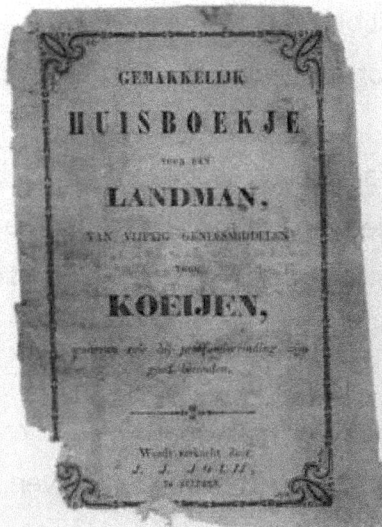

Figure 5. Farmers booklet (Jolij, 1855). Containing 50 remedies (including herbs) for cows, kept by a Dutch farm family that still uses some of these herbs. Dutch cows in a nature reserve area, their herb consumption was studied by the IEZ

Input of ethnoveterinary methodology: a necessity

Ethnoveterinary related activities take place in every society, but in the Netherlands they are undertaken by universities that didn't build up or hire expertise on ethnoveterinary ethics and methodology. There are some disadvantages connected to this situation, as the following examples show.

The agricultural University of the Netherlands in Wageningen (WUR) was asked several times to make inventories of ethnoveterinary ('alternative') remedies used by (mainly organic) farmers. They did this without involving proper ethnoveterinary or related herbal medicine expertise. Four examples, taken out of the resulting reports, are given here:

1.One study [61] mentions the use of "alternative medicine" on 27 out of 30 farms, of which only 1% is phytotherapy. The report lists phytotherapy use without giving a proper definition of phytotherapy. The small amount of herbal remedies (page 27) is mainly due to a lack of knowledge, as many herbal remedies are listed under the "general" category (phytotherapeutic registered remedies like Uterale/Sabine herb and all volatile oil based products of *Mentha, Eucalyptus, Thymus, Melissa* and other species) or as a homeopathic remedy (*Echinacea*).

2. Kijlstra [62] on page 39, mentions a remedy used by goat farmers for wounds "Titrie", without further specification. Any herb expert would have understood that it should be Tea tree (*Melaleuca* spp.) The inclusion of this important, well documented remedy in this way is useless in Kijlstra's report.

3. WUR Researchers asked the advice of a pharmacognosist on what herbal remedies to test to combat *Ascaris suum* (roundworms) in pigs [63]. After a literature search the advice given was (amongst others) Papaya latex, this is a specific product that can be made from unripe fruit or leaves of Papaya tree. For economical reasons Van Krimpen used the Papaya ripe fruit juice (a cheaper product, but without anthelminthic activity) to perform the experiment. The original pharmacognost advice was neglected, and he was not asked to contribute to the publication in the Veterinary Parasitology Journal. The flaws of this research were commented on [64].

The authors wrote that they also tested the herb *Artemisia vulgaris* [63]. However, they referred to literature on *A. brevifolia*. During a presentation on this research in a conference of the Netherlands Association of Phytotherapy (NVF) it became clear that the lead author Van Krimpen could not tell what *Artemisia* species he had used for the experiment (he had ordered *Artemisia*, not specified, from a supplier). As there are many *Artemisia* species in the market, with big differences in application (from kitchen herbs to the effective antimalaria remedy *A. annua*) the species used is an important factor in the study. Although in the Van Krimpen study a beneficial influence of certain herbal products on liver spots was clearly visible, his general conclusion was written as if his results were very negative for the use of herbal products [63].

4. Mul and Reuvekamp [65] published an overview of herbal remedies for poultry worms. With this aim they made an inventory of relevant literature, both ethnoveterinary field work and experimental lab work. The report shows many methodological and linguistic shortcomings. The authors give only 4 literature references in a report mentioning ca 200 plants. Some of the plant names are given with the author who mentions them and the year of publication, but without the actual references. Other plants are mentioned with no indication for a reason. The decision on what to advise their target group (organic farmers) to do regarding future research is based on a kind of non-experimental validation i.e. a judgment (no criteria mentioned) by the two authors with no other experts involved of the (not given) reference publications. Criteria such as 'locally and cheap available and non-toxic' and 'both *in vitro* and *in vivo* data should be available' were used to select a number of plants that should be further examined. Of these 5 plants St John's wort (*Hypericum perforatum*) was excluded because of reported side effects (no references given).

A small part (randomly taken) of the long plant list that forms the body of the publication is copied here:

Echinacae .... Zonnehoed

Embelia schimperi

Emblic myrobalans Ambla

Eucalyptus Gomboom, Gums onbekend nee VIVO olie (waarvan?) Mocsy 1931 AB

Euonymus europaeus Wilde kardinaalsmuts, Spindle

Euonymus verrucosus

The first, third and fourth plant names in this list of six are wrongly spelled and/or not identifiable on a species level. As no references are given by all but one (and this is a very incomplete reference) it remains unclear what the source is of these mistakes.

The fact that nearly all government funded research regarding (both organic and regular) livestock in the Netherlands is performed by WUR-LR (in general, no open competition between research institutes is organized) makes it clear that methodological failures as mentioned above cannot easily be improved.

Participatory research

Participatory research co-operatively documents and validates the ethnoveterinary medicines used by livestock farmers in the developed world. The International Institute of Rural Reconstruction (IIRR) [66; 67] developed the workshop method explained below and it is said to have two major advantages: it reduces the total amount of time needed to develop information materials (a user-friendly manual) and it profits from the expertise and resources of a wide range of participants and their organizations. As such it is an example of co-operative learning or working together to meet common needs to achieve more than one can as an individual. The workshop process results in the recommendation of ethnoveterinary practices and remedies to be included in a manual. The remedies chosen for inclusion are those that can be recommended for use by the general public and farmers to alleviate minor diseases and problems. A similar method was used in a study done in the Yaegl Aboriginal community in New South Wales, Australia, but this study does not include a reference to any specific protocol involved with this methodology, for example the one that has been published for the ethnoveterinary participatory work discussed here that has been done since the 1990s [68]. Both of the references cited for the Yaegl Aboriginal study – UNESCO [69] and Baum et al. [70] contain only a review of the literature on participatory work with references to power and knowledge control between researchers and the researched, lack of benefits given to the researched, the misrepresentation of cultures by researchers, difficulties in attracting funding, biodiversity, resilience, reciprocal rights and responsibilities and aboriginal ethics and not the specific methodology discussed here and used in the Yaegl study.

The use of safe and effective medicinal plants can reduce farmers' input costs, preserve the resource base, enhance biodiversity and protect animal health. If plants are grown on-farm this will enhance the biological interactions on which productive agriculture depends. Results of participatory research can contribute to farm incomes, maintain the resilience of farm communities, promote self-reliance and contribute to an internationally recognized safe and good quality food supply, in addition to providing improved and affordable livestock health care.

It will strengthen rural community capacity building, leadership and skills development. As such it fits the research mandate of serving as a contributor to the economic survival for farming communities. A manual of all validated remedies can serve to preserve the ethnomedicinal heritage of developed world farmers. The manual describes the preparation, use and efficacy of ethnoveterinary medicines in a form that is relevant to livestock farmers, and which can be understood by farmers and the general public [71]. Many European farms that are striving to meet the demands of the niche market for meat products grown without hormones and antibiotics could be helped in a practical way with the use of this method.

Methodology for ethnoveterinary participating research

The research process we outline here, will be described in short and stepwise. More details are to be found in previously published work [45].

1. Compile a list of livestock farmers and other informants.
2. The sample size should be appropriate.
3. All participants should be given consent forms prior to participation in the research.
4. A draft outline of the participant's ethnoveterinary remedies is discussed during a participatory workshop in order to establish that dosages were accurately noted, for input on content, and to clarify any points.
5. The participant-approved drafts are compiled into the draft manual to be discussed on the last few days of the workshop.
6. Medicinal plant specimens are collected where possible and these are identified and deposited as vouchers at the closest herbarium. A voucher herbarium specimen is a pressed plant fully annotated and placed in a Herbarium for future reference.
7. Interview schedules are filled in accordance with published guidelines [72; 73].
- Basic contact data on each informant (phone, email, address).
- The medicinal plants known as well as used (scientific and local names will be recorded).
- From whom, when they had acquired each remedy, their training and background (if any).
- Which plant part used, what stage of growth, collection time, preparation alone or in combination.
- Preparation details: raw, cooked, fermented, purified, dried, ground, pounded, minced,

powdered, washed, bruised, crushed, infusion, decoction, maceration), heated or not. Administration and dosage. Ingested (liquid, dry), chewed, eaten, dropped in or on, sprinkled on, massaged or rubbed in, bandaged on, fumigated. Applied as poultice, soak, wet dressing or bath.

- Interviewers' assessment of the effects of each remedy. Specific pharmacological properties or actions. Specific diseases or conditions (such as lack of appetite).
- Vernacular designations. A disease concept may translate directly (e.g. itch means scabies).
- Frequency, length of therapy, relationship of the initiation of therapy to the stage of disease development, relationship to meals. Precautions to ensure effect or to avoid side effects (e.g. take or do not take on empty stomach, avoid hot food).

Validation of practices

The research team completes the non-experimental validation of the remedies in advance of the workshop. Traditional validation and drug discovery is extremely expensive so a non-experimental method is used [45; 72-75]. This method consists of:

1. obtaining an accurate botanical identification
2. determining whether the folk data can be understood in terms of bioscientific concepts and methods
3. searching the chemical/pharmaceutical/pharmacological literature for the plant's known chemical constituents and to determine the known physiological effects of either the crude plant, related species, or isolated chemical compounds that the plant is known to contain. This information is used to assess whether the plant use is based on empirically verifiable principles or whether symbolic aspects of healing are of greater relevance. For example if the plant is reputed to cause itching or bleeding, the assessment determines if it contains chemicals that can cause itching and bleeding.

Table 1. Non-experimental validation for *Hydrastis canadensis*: Review of the ethnomedicinal literature

Scientific name	Ethnoveterinary use	Published literature	Reference
Hydrastis canadensis	dog bites, bleeding wounds, sprains, post-operative bleeding, deep wounds, abrasions	Berberine inhibited activator protein 1 (AP-1), a key transcription factor in inflammation and carcinogenesis. Berberine has a significant inhibitory effect on lymphocyte transformation, and its anti-inflammatory action may be due to inhibition of DNA synthesis in activated lymphocytes. During platelet activation in response to tissue injury, berberine has a direct effect on several aspects of the inflammatory process. It exhibits dose-dependent inhibition of arachidonic acid release from cell membrane phospholipids, inhibition of thromboxane A2 from platelets, and inhibition of thrombus formation.	76

If ethnobotanical data, phytochemical and pharmacological information supports the folk use of a plant species like (Table 1), it can be grouped into the validation level with the highest degree of confidence. There are four levels of validity [73]:

1. If no information supports the use it indicates that the plant may be inactive.
2. A plant (or closely related species of the same genus), which is used in geographically or

temporally distinct areas in the treatment of similar illnesses, attains the lowest level of validity, if no further phytochemical or pharmacological information validates the popular use. Use in other areas increases the likelihood that the plant is active against the illness.

3. If in addition to the ethnobotanical data, phytochemical or pharmacological information also validates the use in European or North American tradition, the plant may exert a physiological action on the patient and is likely to be effective.

4. If ethnobotanical, phytochemical and pharmacological data supports the folk use of the plant, it is grouped in the highest level of validity and is most likely an effective remedy.

Validation workshop

The workshop that is an essential part of this method involves participatory documentation and validation of the previously recorded ethnoveterinary remedies together with research participants comfortable in talking about their practices (Figure 3). Information on pre-selected ethnoveterinary practices will be prepared in advance for the workshop and working drafts of topics based on each farmer interviews will be given in advance to farmer participants for critical review. This will avoid unnecessary overlap in the participatory workshop. In the workshop the facilitator asks participants very specific questions in a supportive environment about the medicinal plants used [77]. The group interaction is said to produce data and insights that would be less accessible without the interaction, it can refresh memories and encourage sharing [77; 53]. After the first presentations and discussions, the materials are edited, and the first draft published by the workshop staff (Figure 4). This draft is then critiqued and modified by the group. This process can be repeated on each of workshop days. During the last day, the research team discusses the final draft.

Pigs are treated with blended garlic (1 to 5 whole bulbs of garlic per 45 kg of animal in 1 cup of milk) which is put in a rubber tub or trough for pigs to eat. This blend was given once a month from weaning to slaughter and was also given to the sows. Garlic was also added directly to the feed. Pigs were also given 25 kg of mixture made from diatomaceous earth and montmorillonite per 1500 kg feed every day from weaning to slaughter.

Box 1. Ethnoveterinary data collected from Canadian pig farmers

The presentation of ethnoveterinary practices in the manual can follow these guidelines [45; 77]:

At the end of each description there is a number in parentheses with these meanings:
1. Standard veterinary practice or equivalent
2. Traditional practice supported by scientific knowledge
3. Traditional practice that animal healers acknowledge and agree works.
Practices not rated according to 1, 2 or 3, will be rejected.

The final draft of the manual should be ready shortly after the end of the workshop and then published. The manual should consist of easy-to-read, simple information on low-cost, locally available ethnoveterinary practices that can be applied in the developed world.

Ethnoveterinary research as a form of social resilience

Social resilience is concerned with the stability of livelihoods [23]. In the context of ethnoveterinary medicine, social resilience is the ability of farm communities to withstand external shocks (social, economic, political) and stresses like livestock diseases and illnesses. Colonial Americans increased their resilience and lengthened their lives when they bought remedies from native men like Joseph Pye. These remedies then became part of their *cultural capital*. Pye's name was given to the plant *Eupatorium purpureum* which he used to treat typhus, and this plant was official in the U.S. Pharmacopeia from 1820 – 42. Meskwaki medicine was also sold and one medicine man named John McIntosh received $700 to cure a case of dropsy that white doctors had not been able to alleviate [80]. A Sioux doctor named Baptiste treated natives and whites at the Winnebago agency and one native treatment of a Scotsman for a wound in his hand was reported in the *Lancet* [28]. This refutes the assertion by the Dutch government that the potential usefulness of medicinal herbs is unproven. In fact colonial scientists in America conducted many studies on native plants. A liquid extract of cascara sagrada in alcohol was newly added to the pharmaceutical literature in 1898. Dr. J.H. Bundy brought cascara into modern medicine in 1877 and the active principle of cascara was discussed as early as the 1890s. It was recommended for chronic constipation since it was felt that its action was mainly on the lower intestine [81]. All of the pharmaceutical books listed the same uses of cascara as those found among the Native Americans [82]. Other Native American remedies added to the pharmacopoeia from 1890 to 1916 were *Zanthoxylum* species, Oregon grape, *Viburnum* species, *Hydrangea* species and *Pinus* species [28]. Dr. Johann Schöpf listed 335 Native American remedies in his 1787 *Materia Medica Americana*.

Ethnoveterinary medicines can be considered a form of *cultural capital*, which can stimulate livestock production or at minimum provide cost-effective alternatives to allopathic drugs. IEZ asked animal owners what was the source of their remedies. In three cases there was reference to a specific animal Herbal, all three were mid-19th century booklets [83 – 86] (Figures 5 and 6). Since agriculture is a dynamic system, technology needs to keep up with changing consumer demand for food quality and ethnoveterinary medicine can meet some of these expectations. Use of medicinal plants can reduce input costs and enable farmers to stay competitive especially if they are trying to supply the 'chemical-free' niche market with locally grown food products. Increasing farm incomes and enhancing rural communities are strategies that guarantee that there will be farmers in the future- both large-scale farms and more common small-scale diverse farms. If farmers plant and use their own medicinal herbs this will promote environmentally friendly land use and the conservation and sustainable use of biodiversity by the farming community. Sustainable management of farmlands enhances the ecological health and social livability of entire regions.

Conclusion

Veterinary anthropologists have called for more "anthropological veterinary medicine" [49]. They define this as the recognition and utilization of the fact that there are many cultural traditions of striving for and achieving animal health. Scientific medicine is only one alternative and the most funded, there are others deserving of similar support. The chief concern according to veterinary anthropologists must be the provision of animal health, and animal welfare while also producing food and other animal products [49; 87]. The last decades have seen a shift towards intensive animal husbandry that many consumers find unacceptable and that is unaffordable in developing countries. Developing countries do not need to make the same mistakes that countries like the Netherlands made. They can take seriously into account their cultural diversity in healing methods and benefit from the results of a respectful evaluation of local knowledge and traditions [89].

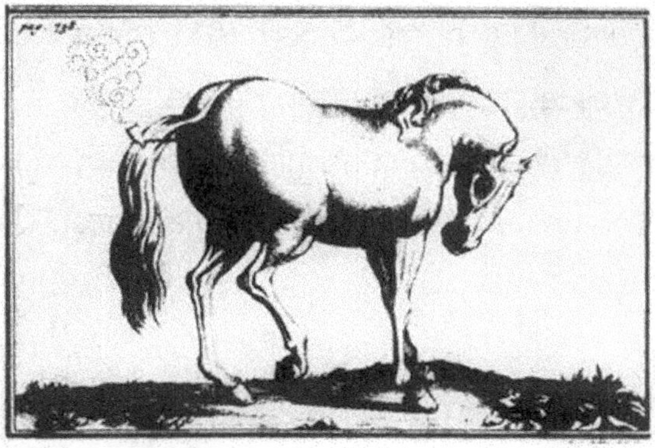

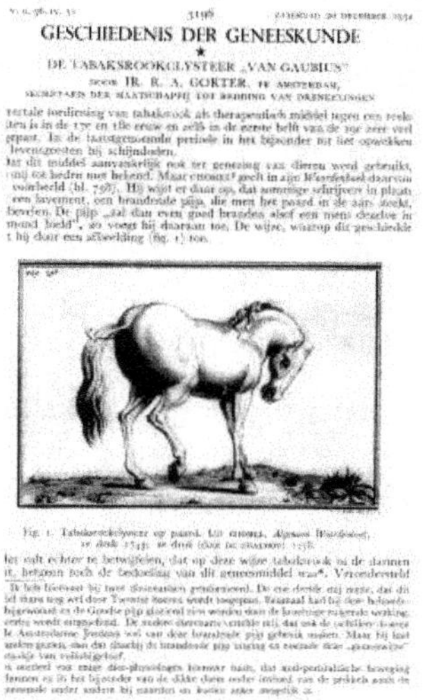

GESCHIEDENIS DER GENEESKUNDE

★

DE TABAKSROOKCLYSTEER „VAN GAUBIUS"

DOOR IR. R. A. GORTER

Figure 6. This illustration from a Dutch 1743 dictionary that goes with the word 'tabaksrookclysteer' shows that the treatment of colic caused by obstipation in horses by introducing the smoke of tobacco (Nicotiniana spp) was in the past common knowledge.

Acknowledgements

Pictures taken by Dr. Evelyn Mathias and Drs A.G.M. (Tedje) van Asseldonk.

References

[1] McCorkle, C.M. 1989. Veterinary anthropology. Human Organisation 48 (2), 156 - 162.

[2] McCorkle, C.M., Mathias-Mundy, E. 1992. Ethnoveterinary medicine in Africa. Africa 62 (1), 59 - 93.

[3] Martin, M., Mathias, E., & McCorkle, C. M. 2001, Ethnoveterinary Medicine: An Annotated Bibliography of Community Animal Healthcare ITDG Publishing, London.

[4] Fink-Gremmels, J. 2005. Toxicology, pharmacology and future directions of JVPT: old and new Paradigms. J. Vet. Pharmacol. Therap. 28, 129–130.

[5] Berkhout, P., van Bruchem C. (eds.). Agricultural Economic Report 2011 of the Netherlands. Report 2011-018. Agricultural Economics Research Institute, the Hague.

[6] Chauvin, C., Madec, F., Guittet, M., Sanders, P. Pharmaco-epidemiology and -economics should be developed more extensively in veterinary medicine. J. vet. Pharmacol. Therap. 25, 455–459.

[7] National Cancer Institute, nd.
   http://www.cancer.gov/cancertopics/pdq/cam/essiac/patient/Page2#Section_24

[8] Zick, S.M., Sen, A., Feng, Y., Green, J., Olatunde, S., Boon, H. 2006. Trial of Essiac to ascertain its effect in women with breast cancer (TEA-BC). J Altern Complement Med. 12, 971-80.

[9] Sillitoe, P. 1998. The development of indigenous knowledge. Current Anthropology 39 (2), 223 - 252.

[10] Peterson, J. 1988. Book review of John van Willigen's Applied Anthropology: An Introduction. American Anthropologist 90, 425 - 426.

[11] Warry, W. 1992. The eleventh thesis: Applied Anthropology as praxis. Human Organization 51 (2), 155 - 163.

[12] Rappaport, R.A. 1993. Distinguished lecture in general anthropology: the anthropology of trouble. American Anthropologist 95 (2), 295 - 303.

[13] Ferradás, C. 1998. Comment. In: Sillitoe, P. 1998. The development of indigenous knowledge. Current Anthropology 39 (2), 223 - 252.

[14] Kielstra, N. 1979. Is useful action research possible?, in G. Huizer and B. Mannheim (eds) The Politics of Anthropology: From Colonialism and Sexism Towards a View From Below, New York, Mouton.

[15] Shore, C., Wright, S. 1996. British anthropology in policy and practice: a review of current work. Human Organisation 55 (4), 475 - 479.

[16] Friedman, John T. (2006). 'Beyond the Post-Structural Impasse in the Anthropology of Development', *Dialectical Anthropology*, Vol. 30, No. 3/4: 201-225.

[17] Said, Edward. 1978. Orientalism. New York: Pantheon.

[18] Latour, Bruno. 1987. Science in Action: How to Follow Scientists and Engineers through Society. Cambridge, MA: Harvard University Press.

[19] Escobar, Arturo 1993 The Limits of Reflexivity: Politics in Anthropology's Post-Writing Culture Era. Review article based on Recapturing Anthropology. Richard G. Fox, ed. Journal of Anthropological Research 49:377-391.

[20] Posey, D.A. 1998. Changing fortunes: biodiversity and peasant livelihood in the Peruvian Andes. Journal of Latin American Studies 30 (3), 682 - 683.

[21] Hastrup, K., Elsass, P. 1990. Anthropological advocacy. Current Anthropology 31 (3), 301 - 311.

[22] Pálsson, G. 1996. Human-environmental relations. In: Descola, P., Pálsson, G. 1996. (Eds). Nature and society: anthropological perspectives. Routledge, London. Pp. 63 - 81.

[23] Adger, W. Neil, 2000. Social and ecological resilience: Are they related? Progress in Human Geography 24 (3), 347 – 364.

[24] Warren, D.M., Slikkerveer, L.J., Brokensha, D.W. (Eds.) 1995. The cultural dimension of development: Indigenous knowledge systems. Intermediate Technology Publications, London.

[25] Milton, K. 1996. Environmentalism and cultural theory: exploring the role of anthropology in environmental discourse. Routledge, London, 266 pp.

[26] Small, E. and P. Catling. 2000. *Canadian Medicinal Crops*. National Research Council Press, Ottawa.

[27] Sproat, Gilbert Malcolm1987. *The Nootka : Scenes and Studies of Savage Life*. Victoria, B.C. : Sono Nis Press. (Originally published as G. M. Sproat, 1868, Scenes and Studies of Savage Life, London, Smith/Elder).

[28] Vogel, Virgil. 1970. *American Indian Medicine*. Norman: University of Oklahoma Press.

[29] Baer, H.A. 1996. Bringing political ecology into Critical Medical Anthropology: a challenge to biocultural approaches. Medical Anthropology 17, 129 - 141.

[30] Baer, H.A. 1997. Introduction to symposium: on-going studies in Critical Medical Anthropology. Social Science and Medicine 44 (10), 1563.

[31] Roling Niels. 2001. From Arena to Interaction: Blind Spot in Actor-Oriented Sociology. admin_en_Roling_long_conference_(2).pdf

[32] Nereu, F., Kock, J., McQueen, Robert J., Scott, John L. 1997. Can action research be made more rigorous in a positivist sense? The contribution of an iterative approach. Journal of Systems and Information Technology 1 (1), 1-24.

[33] Singer, M. 1994. Community - centered praxis: toward an alternative non-dominative applied Anthropology. Human Organization 53 (4), 336 - 344.

[34] Giarelli, G. 1996. Broadening the debate: the Tharaka participatory action research project. Indigenous Knowledge and Development Monitor 4 (2), 19 - 22.

[35] Johannsen, A.M. 1992. Applied anthropology and post-modern ethnography. Human Organization 51 (1), 71 - 81.

[36] Cox, H. 1997. Professional responsibility to the communities in which they work and live. Human Organization 56 (4), 490 - 492.

[37] Hastrup, K., Elsass, P. 1990. Anthropological advocacy. Current Anthropology 31 (3), 301 - 311.

[38] Jiggins, J. 1989. An examination of the impact of colonialism in establishing negative values and attitudes towards indigenous agricultural knowledge. In: Warren, D.M., Slikkerveer, J and Titilola, S. (Eds.). Indigenous Knowledge Systems: Implications for Agriculture and International Development. Studies in Technology and Social Change. No. 11. Ames: Iowa State University, Technology and Social Change Program, pp. 68 - 78.

[39] Katz, C. 1996. The expeditions of conjurers: ethnography, power and pretense. In: Wolf, D.L 1996 (Ed.). Feminist dilemmas in fieldwork, West View Press, Boulder, Colorado, USA.

[40] Casel, Christine K., Brennan, Troyen E. 2007. Managing Medical Resources. Return to the Commons? JAMA 297 (22): 2518-2520.

[41] Lawrence, Peter. A. 2007. The Mismeasurement of Science. Current Biology 17 (15): r583.

[42] Munro, A.J. 2005. Commentary. The conventional wisdom and the activities of the middle range. The British Journal of Radiology, 78 (2005), 381–383.

[43] Aronson, J.K. 2003. Anecdotes as evidence. We need guidelines for reporting anecdotes of suspected adverse drug reactions. BMJ 326:1346.

[44] Moulton, Gary (ed.) 1983. *The Journals of the Lewis and Clark Expedition*. Lincoln: University of Nebraska Press.

[45] Lans, C., Turner, N., Khan, T., Brauer, G., 2007. Ethnoveterinary medicines used to treat endoparasites and stomach problems in pigs and pets in British Columbia, Canada. Vet. Parasitol. 148, 325–340.

[46] Durand, Alfred. 1851. On Hydrastis Canadensis. *The American Journal of Pharmacy*. New Series, Vol. XV11. Ed. William Procter Jr., Philadelphia College of Pharmacy. Philadelphia: Merrihew and Thompson, p. 112.

[47] Sattler, Robert, 1885. Hydrochlorate of Hydrastine. *The Medical News. A Weekly Medical Journal*. Vol XLVI. Jan – June 1885. Philadelphia: Lea Brothers & Co, p. 5.

[48] Pitzer, Geo. C (ed.). 1884. *The American Medical Journal Saint Louis*. Vol XII., 1884, St. Louis, MO: Commercial Printing Company.

[49] Köhler-Rollefson, Ilse and Bräunig, Juliane, 1998. Anthropological Veterinary Medicine: The Need for Indigenizing the Curriculum. Paper presented at the 9th AITVM Conference in Harare, 14th-18th September, 1998.

[50] Srivastava, V.K. 1992 Should Anthropologists Pay Their Respondents? *Anthropology Today* 8 (6): 16-20.

[51] Janssen, Michel. 2002. COI Stories: Explanation and Evidence in the History of Science. Perspectives on Science 10 (4): 457-522.

[52] Bauer, Henry H. 2003. The Progress of Science and Implications for Science Studies and for Science Policy. Perspectives on Science 11 (2): 236-278.

[53] Lans, C. 2012. *Ethnoveterinary medicines used for pets in British Columbia*. 232 pages. Permalink: http://amzn.com/0978346890

[54] Krause G, Zimmermann S, Beutin L. 2005. Investigation of domestic animals and pets as a reservoir for intimin- (eae) gene positive *Escherichia coli* types. Vet Microbiol.106(1-2):87-95.

[55] Campbell, Colin. J. 2002. Petroleum and People. Population and Environment 24 (2): 193-207

[56] Roehr B. 2012. Renewed efforts are needed to curb antibiotic resistance. BMJ. 2012 Nov 15;345:e7778. doi: 10.1136/bmj.e7778.

[57] Overdevest I, Willemsen I, Rijnsburger M, Eustace A, Xu L, Hawkey P, Heck M, Savelkoul P, Vandenbroucke-Grauls C, van der Zwaluw K, Huijsdens X, Kluytmans J.2011. Extended-Spectrum β-Lactamase Genes of *Escherichia coli* in Chicken Meat and Humans, the Netherlands. EID Journal Emerg Infect Dis. 17(7):1216-22.

[58] Vandenbroucke, C. et al, 2011. The Netherlands: can we reconcile veterinarian and human antibiotic use? Poster 3rd world HAI forum: http://www.biomerieux-diagnostics.com/upload/Chritina%20Vandebroucke%20Poster%20Netherlands-1.pdf

[59] EFSA Panel on Biological Hazards (BIOHAZ); Scientific Opinion on the public health risks of bacterial strains producing extended-spectrum β-lactamases and/or AmpC β-lactamases in food and food-producing animals. EFSA Journal 2011;9(8):2322. Available online: www.efsa.europa.eu/efsajournal

[60] Groot, M.J., M.Y. Noordam, G.A. Kleter, A.G.M. van Asseldonk, E. Kleijer-Ligtenberg, S.B.A. Halkes, J. Fink-Gremmels, H.H. van Osch, 2008. Ontwikkelen van fytotherapie als middel bij het reduceren van en/of behandelen van dierziekten (Fyto-V). Rapport 2008.010 RIKILT Wageningen. Stable books in English version available on

[61] Van der Werf, J.T.N.; Kijlstra, A.; Buitendijk, J.; Klink, M.C.M.; Munniksma, K.; Schaaf, R. van der. 2004. Inventarisatie diergeneesmiddelen in de biologische melkveehouderij. *Lelystad : Animal Sciences Group, (Rapport / Animal Sciences Group september 2004).*

[62] Kijlstra, A.; Werf, J.T.N. van der; Buitendijk, J. 2004. Inventarisatie diergeneesmiddelengebruik in de biologische geitenhouderij. *Lelystad : Animal Sciences Group, (Rapport / Animal Sciences Group April 2004)*

[63] Van Krimpen, M.M., Binnendijk, G.P., Borgsteede, F.H., Gaasenbeek, C.P., 2010. Anthelmintic effects of phytogenic feed additives in Ascaris suum inoculated pigs. Vet. Parasitol. 168 (3–4), 269–277.

[64] Lans, C. 2011. Letter to the editor. Validation of ethnoveterinary medicinal treatments. Veterinary Parasitology 178 (2011) 389–390.

[65] Mul, M.F.; Reuvekamp, B.F.J. 2008. Inventarisatie van mogelijke fytotherapeutica met een werking tegen wormen bij pluimvee. *Wageningen: Animal Sciences Group,*

[66] IIRR, 1994. Ethnoveterinary medicine in Asia: An information kit on traditional animal health care practices. 4 Vols. International Institute of Rural Reconstruction, Silang, Cavite, Philippines.

[67] IIRR, 1996. Recording and using indigenous knowledge: A manual. International Institute of Rural Reconstruction, Silang, Cavite, Philippines.

[68] Packer J, Brouwer N, Harrington D, Gaikwad J, Heron R, Yaegl Community Elders, Ranganathan S, Vemulpad S, Jamie J. 2012. An ethnobotanical study of medicinal plants used by the Yaegl Aboriginal community in northern New South Wales, Australia. J Ethnopharmacol. 139(1):244-55.

[69] Baum, F., MacDougall, C., Smith, D., 2006. Participatory action research. Journal of Epidemiology and Community Health 60, 854–857.

[70] UNESCO, 2007. Links between biological and cultural diversity-concepts, methods and experiences - Report of an International Workshop. In: Persic, A., Martin, G. (Eds.), Paris.

[71] Duram, Leslie, A., Larson, Kelli, L. 2001. Agricultural research and alternative farmers' information needs. Professional Geographer 53 (1), 84 – 96.

[72] Croom, E.M. Jr. 1983. Documenting and evaluating herbal remedies. Economic Botany 37 (1), 13 - 27.

[73] Hirschhorn, H. 1983. Constructing a phytotherapeutic concordance based upon Tropical American and Indonesian examples. Journal of Ethnopharmacology 7 (2), 157 - 167.

[74] Browner, C.H., Ortiz de Montellano, B.R., Rubel, A.J., 1988. A methodology for cross-cultural ethnomedical research. Current Anthropology 29, 681 – 702.

[75] Heinrich, M., Rimpler, H., Antonio-Barrerra, N., 1992. Indigenous phytotherapy of gastrointestinal disorders in a lowland Mixe community (Oaxaca, Mexico): Ethnopharmacologic evaluation. Journal of Ethnopharmacology 36, 63 - 80.

[76] Anon, 2000. Berberine monograph. Alternative Medicine Review 5, 175 – 177.

[77] Mundy, Paul and Mathias, Evelyn. 1997. Participatory workshops to produce information materials on ethnoveterinary medicine. Paper presented at the International Conference on Ethnoveterinary Medicine: Alternatives for Livestock Development, Pune, India, 4–6 November 1997.

[78] Smith, Huron, H. 1928. Ethnobotany of the Meskwaki Indians. Bulletin of the Public Museum of the City of Milwaukee. Wisconsin: Cannon.

[79] Barton, Benjamin Smith. 1810. *Collections for An Essay Towards A Materia Medica of the*

United States. Part First, 3rd Ed. Philadelphia: Edward Earle & Co., p. 36.

[80] Pereira, Jon, 1837. *Lectures on Materia Medica, or Pharmacology, and General Therapeutics*. Delivered at the Aldersgate School of Medicine. London Medical Gazette. Saturday September 16, 1837, p. 900.

[81] Erichsen-Brown, Charlotte. 1979. *Use of plants for the past 500 years*. Ontario: Breezy Creeks Press.

[82]Gunther, Erna. 1992. *Ethnobotany of Western Washington: The Knowledge and Use of Indigenous Plants by Native Americans*. Seattle and London: University of Washington Press.

[83] Jolij, J. 1855. Gemakkelijk huisboekje voor den landman van vijftig geneesmiddelen voor koeijen, waarvan vele bij proefondervinding zijn goed bevonden. Zutphen.

[84] Numan, A. 1844. Handboek der genees- en verloskunde van het vee. Utrecht.

[85] Wagenfeld, dr L. 1844. De Vakbekwame Veearts.

[86] Van Asseldonk, Drs. Tedje and Beijer, Helen (2006) Herbal folk remedies for animal health in the Netherlands. In: Ertug, Dr. Z. Füsun (Ed.) *Proceedings of the IVth International Congress of Ethnobotany (ICEB 2005) Istanbul*, pp. 257-263

[87] McCorkle, C.M. 1995. Back to the future: Lessons from ethnoveterinary RD&E for studying and applying local knowledge. Agric Human Values 12 (2): 52 – 80.

[89] TAHCC. Alternative Animal Health Care in British Columbia: a manual of traditional practices used by herbalists, veterinarian, farmers and animal caretakers. The Traditional Animal Health Care Collaborative. Victoria, British Columbia, Canada, 2004.

The moral case for animal welfare for snakes and alligators

Abstract

In the 1990a the struggle over the proper use of the Nariva Swamp – as a park or for rice production was framed in terms of co-management. At that time conservation and preservation approaches were being challenged academically and were even criticized by Hanna Siurua in 2006 as "fortress conservation." However Alex Clapp acknowledges in 2004 that not all communities are coherent, united, or strongly attached to place. This means that co-management is not always possible. This paper discusses co-management from the viewpoints of animal welfare and ethics. The core of the issue is that large and potentially dangerous species such as snakes and caimans which depend on the Nariva Swamp cannot be co-managed. Specific animals and their treatment in terms of welfare and ethics are also discussed: these are Happy Feet the Emperor penguin, Elvis and Eric, two Australian crocodiles and several nameless snakes.

Keywords: Trinidad and Tobago, Nariva Swamp, snakes, caimans, alligators, animal welfare

Animal welfare for snakes and alligators?

Introduction

In the past sociologists assumed humans had exceptional traits that freed them from the constraints and limits imposed by Nature (Murphy, 1995; Catton and Dunlap 1978; Dunlap and Catton, 1979). Sociologists tried to analyze man's ascent from animality and embeddedness in nature, reinforcing anthropocentrism and self-aggrandizing the social while ignoring the ecosystem-dependence of all species including man. Today Environmental Sociology investigates the social dimensions of environmental problems, including the conduct of environmental politics, processes of making environmental claims and the construction of environmental knowledge (Murphy, 1995 Murdoch, 2001).

Murdoch (2001) questions environmental sociology's continued attachment to the 'social', and the continued division between society and nature which is inhibiting the emergence of a new and genuinely 'green' sub-discipline. Brydon (2006) wonders if anthropology should abandon Nature altogether rather than move to the myth that natives have always lived more harmoniously with nature than non-natives (even in the face of empirical studies of disharmony, error, and ecological destruction). Some environmental degradation is poverty driven, but environmentally damaging behaviour also results from

gender interests, "striking bargains with the patriarchy' and ideologies (Jackson, 1993; Cornwall, 1998).

Political ecologists and critical geographers see themselves as guardians of traditional spaces and prior

uses. Within this ethical framework, conservation programs are examined not for what they could

accomplish in protecting endangered species or ecosystems, but on their human impact (Clapp, 2004;

Castree, 1999). Tompkins et al. (2002) claim that participatory approaches do not necessarily produce

sustainable utilisation of resources. They write that there are limits to co-management based on the

complexity of the resource, the size of the user pool, the reality of the social construction of networks and

spaces in which co-management takes place and the geographical area.

In this paper I will discuss the struggle over the Nariva Swamp in Trinidad (Figure 1) in the 1990s. The

Swamp is the natural habitat of snakes and caimans (Family Alligatoridae Order Crocodylia, *Caiman*

crocodilus) (Figure 2).

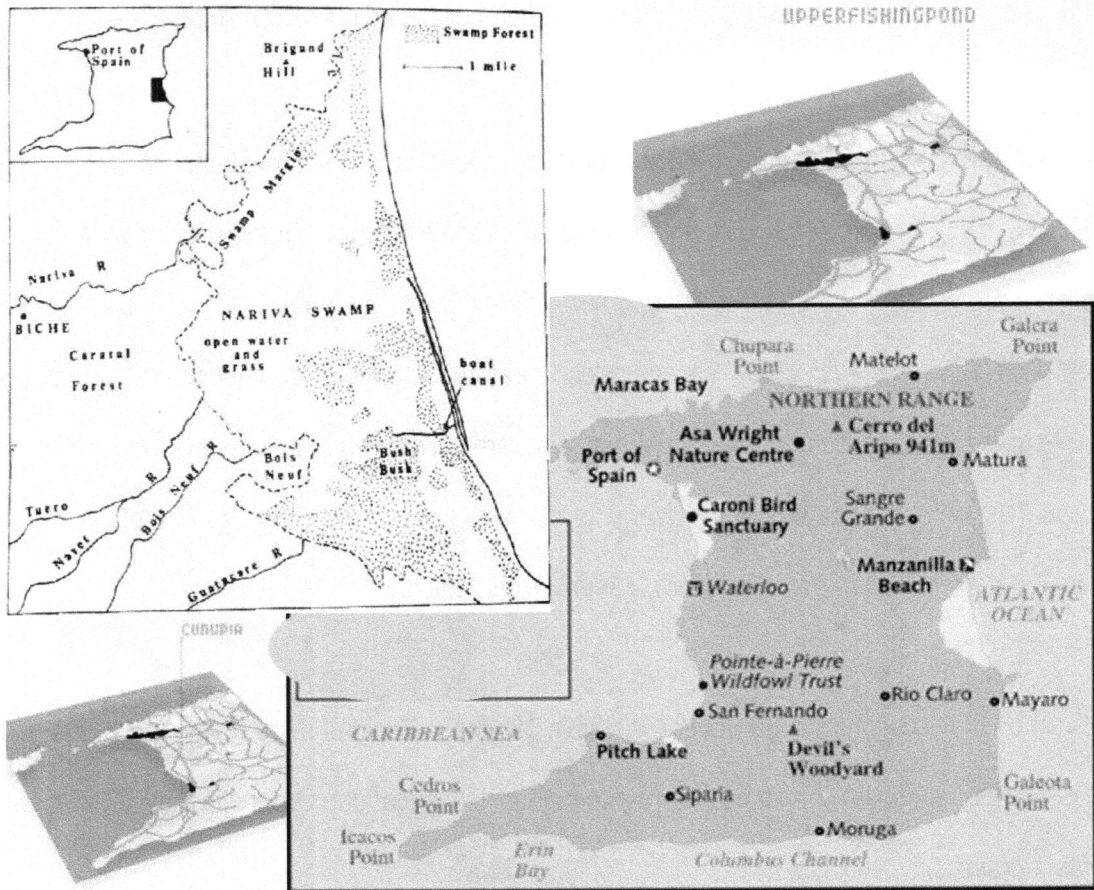

Figure 1. The location of Nariva Swamp in Trinidad also showing other area of interest (sources Aitken, 1973, wikispaces, nona.net)

Figure 2. Illustration of a Caiman crocodilus and an Anilius scytale (1701–1705) by Maria Sibylla Merian. Colored copper engraving from Metamorphosis insectorum Surinamensium II.

The environmental framing never included animal welfare; the terms of the debate were rice production versus national parks in a country that favoured development. Internal and external researchers and stakeholders attempted to resolve the dispute with the new environmental co-management discourse. For example Keeler and Pemberton (1996) claimed that "one of the positive features of this situation [conflict in Nariva Swamp] in that both local and international environmental groups are firmly behind the idea of sustainable use by Nariva by people, including its use for agriculture. There is very little sentiment or rhetoric for simply making Nariva into a park or denying all uses except ecotourism. This attitude makes sustainable consensus solutions more feasible."

I will discuss co-management from the viewpoints of animal welfare and ethics; the core of the issue is that large and potentially dangerous species such as snakes and caimans cannot be co-managed. The largest snake recorded in Trinidad was found on the Cocal coconut estate on the boundary of the Nariva swamp, it was an anaconda *(Eunectes murinus gigas)* 9.1 metres long (Anon, 2009).

Multitudes of mapepires

Most of the snakes in Trinidad are called mapepire, an Amerindian-derived word (Boos, 2001). The species dependent on Trinidad's wetlands are fish-eating water snakes (water mapepire *Helicops angulatus* (Colubridae), huille *Eunectes murinus gigas* (Boidae) (huille comes from the Amerindian word ioulia meaning another black and yellow snake according to Boos, 2001), mangrove mapepire *Liophis cobella cobellus* (Colubridae), water coral *Hydrops triangularis neglectus* (Colubridae) and the *Cook's tree boa Corallus hortulanus cookie* (Boidae). The macajuel *Boa constrictor constrictor* (Boidae), the mapepire balsin *(Bothrops atrox)* (Viperidae), the mapepire z'ananas *Lachesis muta muta* (Viperidae), the aggressive machete savane *Chironius carinatus carinatus* (Colubridae), the false mapepire *Leptodeira annulata ashmeadi* (Dipsadidae), a constrictor called ratonel *Pseudoboa neuwiedii* (Colubridae) and the horsewhip *Oxybelis aeneu* (Colubridae) live in a variety of habitats close to water, including the Swamp. The checkerbelly snake *(Siphlophis cervinus)* (Colubridae) is rare. Swamp-based snakes can tolerate a salinity of 12 ppt (seawater is 36 parts per thousand (ppt)) (Murphy, 2011). Other species found in the swamp are the zandolie lizard *(Ameiva ameiva)*, the matte *Tupinambis teguisin* (Teiidae), the gecko, *Thecadactylus rapicauda* (Gekkonidae) and the paradoxical frog (the tadpole is bigger than the adult frog) *Pseudis paradoxa caribensis,* the rare manatee *(Trichesus manatus)* and the rare otter *(Lutra enudris)*.

The Nariva Swamp (6,234 hectares) is the largest wetland in Trinidad and Tobago and is located on the east coast (1994 - 1996 White Paper on Agriculture, page 3). The other wetlands are the Caroni Swamp (5,611 hectares), South Oropouche Lagoon (5,642 hectares) and Fishing Pond (1,220 hectares). The wetlands include mangrove forests, manmade lakes and inland savannahs, such as Erin (Oropouche Lagoon 40 hectares) and Aripo (1800 hectares, bordering Nariva Swamp). Caroni was the first protected wetland; other swamps were converted into rice production as early as 1917 when the British Colonial Government established the Plum Mitan rice scheme (1918), and the Fishing Pond (1954) and Oropouche Lagoon schemes (1917) (Taitt, 1999). The Nariva Swamp became a forest reserve in 1954. The Bush Bush section of the Nariva Swamp (1408 hectares) is an area of high ground that was declared as a wildlife sanctuary in 1968, and a prohibited area in 1989) (Figure 3). Two illegally squatting rice farmers filed a constitutional motion against this legal notice of the state re "the right to enjoyment of property" (Wildlife Section, 1993).

The Nariva Swamp has been threatened in the past by several illegal activities; squatting, cannabis and rice farming, grazing livestock, overfishing, timber harvesting, hunting and excessive trapping of birds for the pet trade (Ramsar, 1996). People living in the Kernahan and Cascadoux communities located on the borders of the Swamp used birds, cascadura, conch, snakes, agouti, caiman and crabs from the area. Some had legally owned lands in nearby Manzanilla, but squatted illegally in the Swamp. The Wildlife Section (1993) documented that these communities illegally sold birds to make a living and blue and gold macaws (*Ara ararauna*) were wiped out by the pet trade. The *Ara manilata* (red-bellied macaw) was threatened due to habitat destruction of the palm (*Roystonea oleracea*) (Wildlife Section, 1993; Bonadie and Bacon, 2000; Duguid et al., 1996; Nathai-Gyan, 1997). This particular set of small-scale squatters had not been in the Swamp before 1930 (Sletto community map from 1997). Twenty of the 61 households in Kernahan were established in the 1970s to create a "new Penal" (East Indian settlement) (Sletto, 2002).

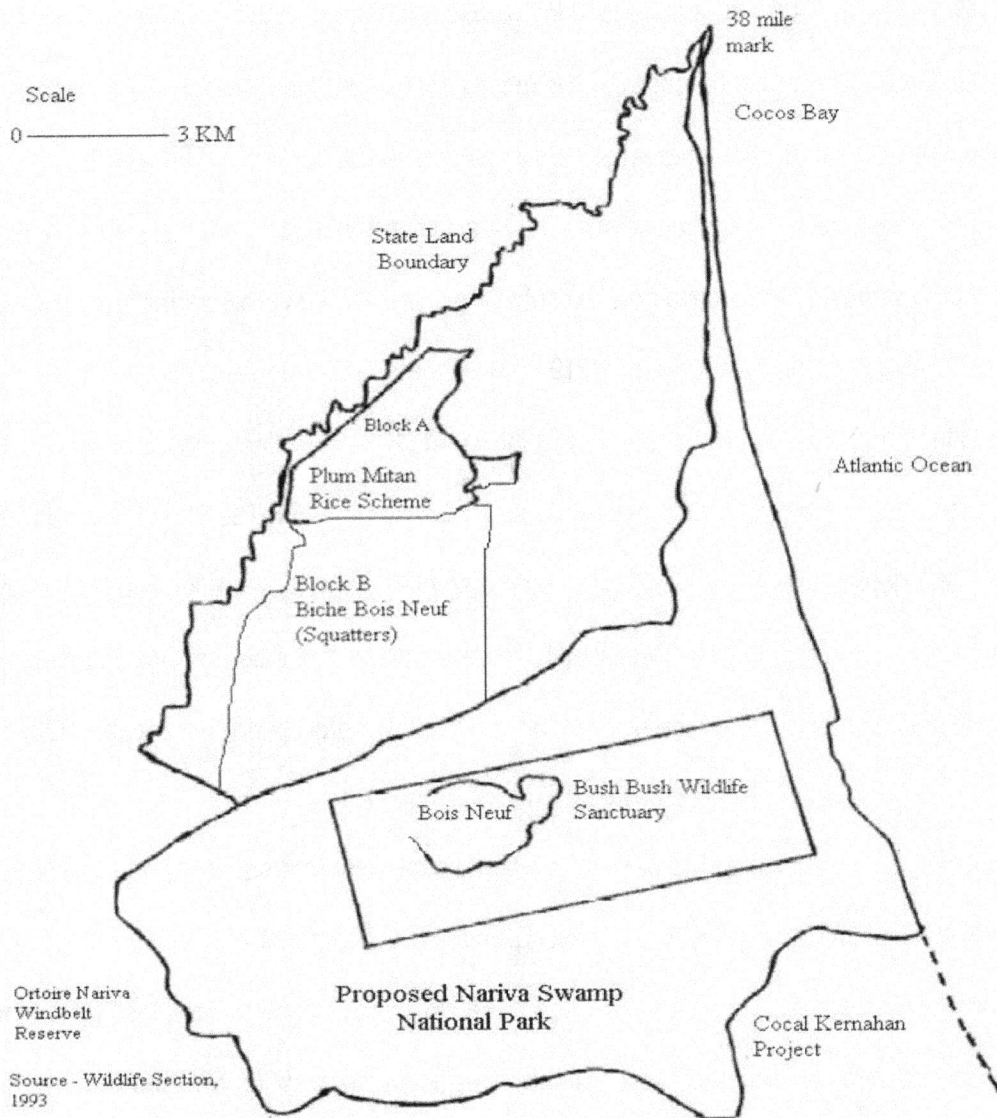

Figure 3. Location of major sectors in the Nariva Swamp (source Wildlife Section, 1993)

Methods

In the 1990s I went to the Swamp weekly to conduct two projects. The new co-management environmental

discourse limited "stakeholders" to those physically present. I was an NGO-consultant setting up an

ecotourism venture and I conducted a gender-based analysis on governance aspects of the Nariva Swamp.

I also paid one of the ecotourism participants to collect data on medicinal plants for my Masters thesis. Due to these efforts I was listed as an "external user" of the Swamp (Kacal, 2000).

Keeler and Pemberton were co-creators of an interdisciplinary research team of University of the West Indies and University of Georgia scientists established to conduct research on the sustainable development of the Nariva Swamp. Their objectives were to promote wise use of the Nariva Swamp, to improve the welfare of the Kernahan community and the wider society from the use of Nariva's resources and to contribute to UWI teaching and research. Basically the project called for economic development that would have required a lot of professional management expertise for its continuance – professional job creation.

Left out of this project, the Centre for Gender and Development Studies obtained funding from CIDA and other sources and conducted a gender sensitive research project in the swamp (the project that I was involved in) (Hardjoeno et al., 1996). The Nariva human community was studied by young participant researchers (Cross et al., 1999; Durbal, 2000; Hosein and Cross, 2000). They used interviews, participant observation, ethnography and participatory approaches that included workshops on time lines, resource use charts and community and benefit flow charts. The governance aspect that I conducted examined the history of governance and policy issues related to the Nariva wetlands and surrounding areas and the countervailing interests of various stakeholders. It included an evaluation of state policy toward agricultural resource use and settlement in the Nariva area (e.g. agricultural policy, infrastructural development, political context, allocation of land rights).

Open-ended interviews were held with the Nariva stakeholders listed below from July to September 1999. One current and one former staff member of Wildlife Section, Forestry Division, MALMR. Three former Ministers of Agriculture/Environment - their periods in office were (1950s to 1985), (1986 – 1991) and (1995 – 2000). Two members of the Pointe á Pierre Wildfowl Trust, three members of one rice farming family, a zoologist from the University of the West Indies (UWI), a member of the Caribbean Forestry Conservation Association (CFCA), a former junior Minister of Environment, a former head of the Trinidad and Tobago Field Naturalists Club, a staff member of the National Flour Mills, a former member of the Agricultural Development Bank (ADB), a member of the Environmental Management Authority (EMA), an ecotourism operator, an environmental lawyer and three agricultural consultants. The secondary literature on rice production, gender analysis, local environmental issues, governance issues and all previous documents related to the Nariva Swamp, from the UWI library and the library of the Ministry of Agriculture at St. Clair was reviewed.

Results

Nariva Swamp conflict and governance

In 1992, the State Lands Division of the Ministry of Agriculture felt that it was their mandate to accept applications for State Lands within the Plum Mitan area, which led to the land grabbing situation in the Nariva Swamp. The Oropouche, Caroni and Fishing Pond swamps had been damaged by saltwater intrusion linked to previous rice production (an estimated $2 million dollars worth of damage), the farmers responsible then moved into Nariva (Wildlife Section, 1993; Planning Associates, 1981). Squatting, depletion and moving (shifting cultivation) is attributed to ancient Spanish land laws which were unchanged during British colonization from 1802 to 1962 and to the behaviour of early sugarcane planters (Wildlife Section, 1993; Driver, 2001). This is an example of how the colonial past continues to organize experiences in the present (buried epistemologies) (Willems-Braun, 1997).

The official chain of management of the Nariva Swamp is provided in Figure 4. Many of the activists working to save the Nariva Swamp were female. National Parks and Wildlife had a minimal budget ($40,000 one year, one jeep, personal funds for gas) (Bacon et al. 1979, p.192), no substantial posts and a long chain of command. It was seen as a "softer" area best left to female "troublemakers" with Biology degrees as opposed to the "male bastion" of Forestry. In contrast the wage expenditure for one rice project (Plum Mitan 1 in 1984) was more than the Forestry Division's 1992 Allocation.

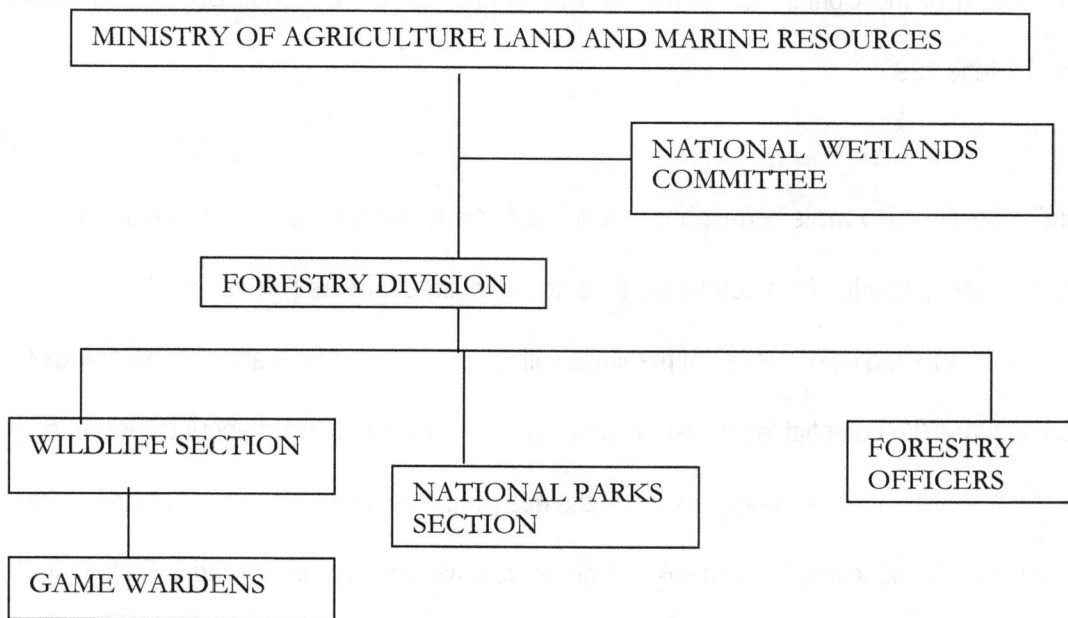

Figure 4. Existing institutional and administrative framework for Nariva Swamp at the time of the research study (source: IMA, 1999)

Forestry officers thought that the latest World Bank National Parks plan would reduce their professional territory: Even though a National Parks policy was approved in principle (Kacal and Homer, 1993) the Wildlife Section could not stop small and large-scale rice farmers from illegally squatting in the swamp with oral permission from certain Ministry of Agriculture officials. So the Wildlife Section created a multi-pronged strategy, one component of which was to approach the Convention on Wetlands (Ramsar, Iran, 1971) for help since the Government had already shown that it would take international obligations like CITES seriously. This intergovernmental treaty involves a commitment of its member countries to plan for the "wise use", or sustainable use, of all of the wetlands in their territories.

Once admitted to the Ramsar network, the Wildlife Section gained access to the international resources they considered necessary for the survival of the Nariva Swamp, and they were able to attend the Kushiro conference [5th Meeting of the Conference of the Ramsar Convention on Wetlands Contracting Parties Kushiro, Japan 9-16 June 1993],

The Wildlife Section also wrote correspondence to the Ministry trying to get the issue resolved. They collected the evidence needed for a court case to remove the large farmers (Wildlife Section Historical Issues Document) and held a series of meetings with judges Stewart Best and Nolan Bereaux to discuss this document and the potential legal case. A large public outcry was created about the issue to force the hands of the Minister of Agriculture; this made the rice policy less politically feasible than before. Powerful organisations in Trinidad and Tobago linked together to have the issue discussed in Parliament.

Ministry of Agriculture support for rice farming

A vertical social capital exchange network between Agriculture Ministers and certain large-scale rice farmers allowed them to access to the natural resource base (free land and water, paid no taxes) in Nariva Swamp, which was far from their home locality of Cunupia and Las Lomas. The social capital network included village labour, mechanised technologies, subsidies for rice production (one former poultry farmer paid off his production debts), a refurbished mill and access to government policy and plans (for example the Draft Second Five - Year Plan 1964 – 1968; MALMR, 1983; MALMR, 1985; FAO, 1957; Lee and Jacque, 1993; 1994- Draft. White Paper; NEDECO, 1981; NEDECO, 1983; OCTA, 1970; Review of the Agriculture sector (1993), 1995; Seereeram et al., 1991; White paper on Agriculture, 1979). In what Bourdieu (1977) describes as 'officialising strategies', the particular interests of key sectors of the rice farming community became identified with the general interest need for national food self sufficiency.

Three Ministers of Agriculture tried to steer a middle course that would appease both the environmentalists and the rice farmers (interview extracts below):

> the large rice farmers gained access to rice land because [this Minister] made a cook in the swamp, and gave them permission."

> "The Bearded Akaloo wanted to do a project in the rice area. Ministers don't give out land, but I offered to see what I could do, and he applied for 200 acres. The Ministry of Agriculture offered 100 acres within the area abandoned within Plum Mitan. The Rice mill was refurbished in my time and the 89 cents subsidy was given in my time. Guyana and Suriname rice was cheaper but it was a deliberate policy to get more local rice, to support local people and the big Akaloos. What they did since then I don't know. The only person I dealt with was Zahir Akaloo, he had a wonderful plan, it benefited the Ministry, he was a leader in field of rice cultivation; Plum Mitan developed under him."

One former Minister of Agriculture touched on the conflict of interest involved in a State institution giving loans to the large-scale rice farmers to grow rice on land to which they had no legal tenure, by claiming that the ADB staff approved the loans for a different area. Three of the five families farming illegally on 1200 hectares got $7 million per year in rice subsidies. Production increased from 91.5 tonnes of paddy rice in 1986 to 4377.9 tonnes in 1992 (TT$12 million production value) (IMA, 1998; Institute of Marine Affairs, 1999).

After the Ramsar agreement

Trinidad and Tobago became a Contracting Party to the RAMSAR Convention in December 1992, with effect from April 1993, and the Government eventually appointed a National Wetlands Committee in 1995 to write the country's first wetlands policy. Members came from the San Juan Rotary Club, the Pointe à Pierre Wildfowl Trust, an NGO-person, from UWI and from the IMA. The two Ministry appointees did not attend meetings. The Ramsar assessment pushed for squatter regularisation and an EIA. The required Environmental Impact Assessment was conducted by the Institute of Marine Affairs (IMA, 1998; Institute of Marine Affairs, 1999) (Figure 5). The EIA found that prior to large scale farming agricultural activity in Block B was 110 hectares out of 6000 hectares for Nariva Swamp. The percentage of the area affected was 1.8%. The social non-user value of $608,000,000 at 1.8% of the swamp = $10.9 million. For Large Scale Rice Growing of 1,200 hectares or 18.7% plus previous agricultural activity displaced, or 1090 hectares Environmental damage = $110.5 million.

The Wetlands Committee pushed for Nariva to become an official Ramsar site or *Wetland of International Importance* as a foundation for the future, State-owned Caroni Ltd had 20234.3hectares of sugar, citrus and rice producing lands and some of the large farmers were allowed to farm there in February 1998 after the Nariva Swamp Environmental Impact Assessment was completed. Theresa Akaloo said that the 500 acres given to them was not enough… [she said that] the survival of Nariva is not incompatible with the growing of rice in the wetlands and that the farmers were not against preserving the swamp (Rostant, 1998). The small farmers could have been allowed to lease Caroni lands as well. Instead in June 2000, 23 Kernahan residents who paid the surveying and legal fees obtained a 30 year lease on 3.5 acre blocks of land in

Block E under the Ministry of Agriculture's Accelerated Land Distribution Programme for which they were to

pay $90/rent a year (Doodhai, 2000). The government then put in place some infrastructure (roads,

electricity, and water).

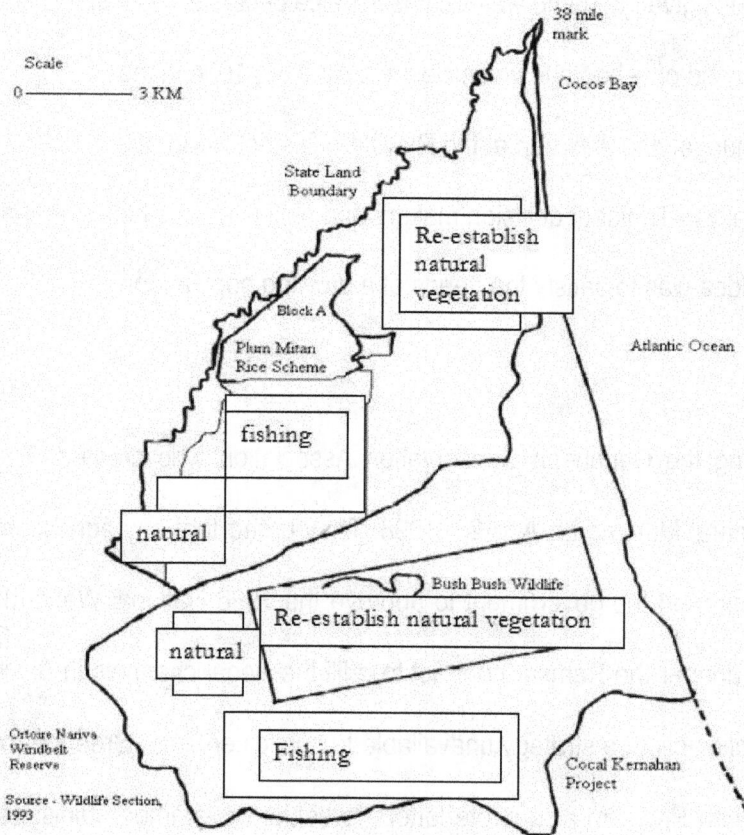

Figure 5. Proposed Management of the Nariva Swamp (IMA, 1999)

The conflict over which discourse would prevail, food security vs. environmental conservation, was

played out in the Magistrate's courts:

"Magistrate Kenny Persad talked about men who should be complimented for growing food instead of being charged for damage to 2 parrots and 4 monkeys, and a minimal fine of $43 was given to the large-scale farmers who bulldozed the swamp. All the time, the Swamp and the environment were not considered important. The Wildlife Section was asked why we were harassing little people. Magistrate Jurity talked about 2 birds in a cage versus agricultural development. There was also the claim that the laws were white men's laws. The police as well were not sympathetic, and wouldn't give the game wardens assistance because they said they had more important things to do and only had a few vehicles. On April 30, 1996 in the court matter vs. Jabar, Sukhoo, Jaimungal and Jaikaran at the Rio Claro District Court, the total fines were $5,100.00. Traditionally in Trinidad, decision makers and policy makers have looked on swamps as useless. The idea was to satisfy the needs of a growing population. How can we use the useless land?"

Advocacy also took place through enrolling the Caribbean Conservation Association who created "The Nariva Resolution" at the 32nd Annual General Meeting on Aug 28, 1998. They urged that any agriculture should be confined to Block A. They encouraged the government to approve the 1996 National Wetlands Policy and called upon the government to honour the Ramsar contract to refill the illegal channels in Block B. Molly Gaskin took on a campaign of public action, a strategy unavailable to public servants. Greenpeace and the WWF Latin American and Caribbean Program also wrote letters to senior government ministers, More than 22,000 people petitioned the Government to protect the Nariva Swamp.

Animal welfare – largely absent

Research on and commitment to animal welfare is fairly recent. Research on the sense of fairness and justice in capuchin monkeys and other primates was reported on in the Wall Street Journal (Begley, 2006), in the Journal Nature (Brosnan and De Waal, 2003) and on the BBC webpage (http://news.bbc.co.uk/1/hi/sci/tech/3116678.stm). While the value of 2 parrots and 4 monkeys was being dismissed in the Magistrate's Courts there was no concerted effort to discuss the animal welfare issues of the Swamp's animal inhabitants by environmentalists. Part of the problem was the non-cuddly nature of the animals involved – caimans and snakes such as *Hydrodynastes gigas* and *Lachesis muta*.

Another stumbling block was the effort of the female professionals to avoid making green issues into emotional issues (and thus less rational than white men and closer to animals and Nature Bailey, 2005); and they prided themselves on their "objective rankings" of the top three sites for national parks, which did not include all of Nariva… "the Thelen and Faizool draft policy for national parks was approved in principle. Sixty-one national parks was considered too many, but Bush Bush was one [of those considered important] (Figure 3). Sylvia Kacal (1997) stated at the Nariva Swamp Seminar held at the University of the West Indies in May 1996 "it is clear that small island states cannot set aside many large wilderness areas for strict preservation." Large areas have been set aside for industrial parks like Point Lisas and La Brea. Male interviewees accused prominent activists as being "emotional Gaskinites" and "Wildlife was always getting itself hurt" [Over the disappointing manatee project. The San Juan Rotary Club not only took all the credit, but also restricted access to people they knew].

Community conservation

Hanna Siurua (2006) preferred community conservation models because "communities down the millennia have developed elaborate rituals and practices to limit offtake levels, restrict access to critical resources, and distribute harvests". Siurua acknowledged that certain species (tigers and rhinos) could not be co-managed. Contrary to Siurua's claims, Worth (1967) found that local hunters in the Nariva Swamp did not know one rat from another (*Zygodontomys brevicauda, Oryzomys laticeps* and *Heteromys anomalus*) and had knowledge only of the agoutis and larger game animals (*Cuniculus paca, Dasypus novemcinctus, Didelphis marsupialis, Mazama americana* or *Dasyprocta fuliginosa*). They also did not know that there were two different kinds of murine opossums (*Marmosa* sp.), not just young and adults. Worth (1967) himself naively describes the "progressive decline" of the Bush Bush rodents (Figure 3) after he started to trap them and amputate their toes in order to identify them; not acknowledging that his human presence and interference may have contributed to their decline. He writes that he "fails to weep" for them and he moves on to trap and band birds.

Mahy (1997) and James (1994) correctly identify the co-management of the marine turtles on the Matura beach on the north-coast of Trinidad as a successful example of community conservation. However leatherback turtles only visit the island for a specific reproductive period. A national park with a constant stream of visitors would have provided the jobs the community needed and encouraged the protection of the wildlife. Small and large farmers fought over the truck-borne water supply (Cross et al., 1999) and did not collaborate in other areas (Hosein and Cross, 2000). Community members told researchers that their survival was more important than that of the snakes and alligators that threatened them. Their livelihood also took precedence over having a balanced environment and one farmer had illegally cut as much as 35 acres of land for cultivation (Cross et al., 1999). In the middle of my own ecotourism training project we were met with a freshly killed snake and caiman left in the road for everyone to see. Ecotourists still see dead snakes in Nariva (a dead water coral (*Hydrops triangularis neglectus* T&T Feb08 TRep (John)).

There are many examples of non-environmental attitudes in the Masters thesis of Amar Wahab and in his 1997 publication and in Durbal (2000). These interviews clearly show the priorities of the swamp communities: four of the men and women interviewed said that any management plan for the swamp should share lands, with a minimum of 10 acres per family. One man from Brigand Hill wanted 100 acres. Wahab's interview with the Plum Mitan Women's Group reveals that the large farmers opened their eyes to large-scale rice production. The Biche male interviewees and a Biche family praised the large farmers for improving the access to fish by pumping out the water from their rice fields. Only one person of 40 interviewed, a woman from Kernahan, said that the Swamp was beautiful.

Discussion

Co-management discourses

Kalema-Zikusoka (2005) points out that it is the "problem" wildlife that reduce the value of the surrounding land which then attracts poor people and squatters to that particular area (like Nariva Swamp) which has no competing human users. Driver (2001) quotes reports stating that it is the avoidance of crop theft that drives squatters to steep inaccessible hillsides; they have no socio-cultural value. If the eastern side of Trinidad remained undeveloped that would allow the animal populations space to regenerate. Trinidad has forty-seven snake species. The pit viper *Bothrops asper* is said to have one of the smallest home ranges (3.71 ha to 5.95 ha) (Sasa et al., 2009). Ruschenberger's Treeboa, *Corallus ruschenberger* was found in the Caroni Swamp (Figure 1) at a density of 11 snakes per square kilometres (Taylor et al., 2011). The huille *Eunectes murinus gigas* is one of the largest snakes in the world and they risk death if their swamp habitat dries out for too long (Barrio-Amorós and Manrique, 2007). The existing nature reserves such as Asa Wright (526 hectares) cannot provide sufficient space (Figure 1).

When environmentalists claim to be safeguarding remnants of pristine nature, social science critiques depict them as evicting people and erasing cultural landscapes in order to reach an unattainable ideal of wilderness (Clapp, 2004; Driver, 2001; Leach and Fairhead, 2001). Driver (2001) claims that "science-policy processes have created and sustained an environmental category: the degraded landscape" ignoring landscapes that would not exist without continuous human intervention. Driver does not acknowledge the lived experience of the "bourgeois" environmental critics and prefers to claim that the landscape is not "pristine"; further changes should therefore be unremarked upon. Seemingly ignored by external researchers were the consumption studies by Daniel Miller (1994) which stated clearly how devoted Trinidadians are to modernity, consumption and material objects – a culture seemingly ill suited to co-management. Sletto (2002) did not have a co-management agenda and saw accurately the social construction of the "noble swamp dweller," who knows and wisely manages the resource.

It is my contention that social science policies like "wise use" are more likely to produce jobs and research opportunities for social scientists (facilitation, management etc.) than "unused nature" and freshwater wetlands and that is the reason for their popularity. Jacques (2006) claims that some proponents of "wise use" are actually environmental sceptics. Sceptics believe that the domination of nature through technology is necessary for successful modernity, progress and affluence. Sceptics assert that there are no environmental problems that threaten environmental sustainability, and that the environmental movement is obstructing human progress (arguments also heard in Trinidad and Tobago).

Attitudes to wildlife

Kupfer (2003) claims that attitudes and interests sometimes determine the aesthetic appreciation people have of Nature. Non-competitive individuals may approach a steep slope as exploratory rather than as confrontational, or as a personal challenge. It is impossible to live in Trinidad and expect that wood slave geckos (*Hemidactylus mabouia*) will not be regularly visiting your house eating the insects that are also visiting. Attitudes to larger, less ubiquitous animals are different. An environmentalist once claimed that Trinidadians divide animals into two categories those that bite and those that "bite nice" (are good to eat). Boos (2001) wrote that the most common excuse that he heard for killing snakes was "I didn't know if it was poisonous". Trinidad has four species of venomous snakes, Tobago has none.

There have been no anaconda-related human deaths in Trinidad but a pet dog was found in the stomach of one (Boos, 2001). Below are five news stories showing excerpts of snakes and caimans coming into contact with humans, with no reports of harm:

1. A villager from Debe (on the West side of Trinidad) complained about poor drainage leading to an influx of reptiles. "Caimans, he said, hide under vehicles. We are afraid of being bitten by these snakes and caimans. The caimans are especially bold and the bigger ones will stand their ground when you confront them" (Asson, 2010).

2. While Met Office officials are hopeful the Caroni River will not burst its banks, they have advised residents in low lying areas to take all possible precautions. ...there are caimans in the water which have started eating animals," Bamboo resident Premnath Bisnath said (Paul, 2011).

3. The Cipero River broke its banks and ..set off massive flooding... Mechanic Andre Gupta, 34, said the flood brought a caiman into his house. "I lost everything," he said (Asson, 2010).

4. Residents at Roberts Street and Emmanuel Trace complained yesterday that former Caroni lands which were converted into rice fields were covered with water, and this resulted in several homes

being flooded. They added that ,,, several reptiles including caiman and snakes entered their homes, and they were forced to get rid of the unwanted reptiles with no assistance from the relevant authorities (Anon, 2010).

5. Apart from the mounds of rubbish covering the land, residents said that the water-logged nature of the lot of land is a haven for scorpions, rats, caimans and snakes. "Forget about paying to go to the zoo, come here in Aripero and you can see snakes and caimans for free," was the sarcastic view of a resident regarding the land (Anonymous, 2010).

Two naturalists recorded nine encounters with the mapepire balsin (*Botrops atrox*) that occurred within the rural Talparo area between 16 February and 8 April, 2009. Three people were bitten and survived after being treated with antivenin (Quesnel and Abdullah, 2009). Five small snakes were found in Quesnel's neighbour's hedge after he was bitten. Talparo is sparsely populated area with cocoa and coffee plantations and two rivers and so is relatively safe for snakes, but has enough people to attract rodents. The authors claimed that no one had ever been bitten by a mapepire on a Field Naturalist hike but recalled one occasion in which 89 people walked in single file passed a seemingly sleeping full-grown mapepire balsin before the 90th person saw it, screamed and stopped.

Animal welfare for food and wild animals

Wolfe (1998) points out that several studies in ethology, field ecology, cognitive ethology, and language experiments over the past twenty years have shown that animals have the properties formerly thought to be distinctly human-- language, tool use, tool making, social behaviour, altruism, non-verbal language, a hatred of boredom, an intelligent curiosity towards their environment, love for their offspring, fear of attack, deep friendships, a horror of dismemberment, a repertoire of emotions and the same capacity for exploitive violence as humans, self-awareness, the ability to engage in both deceptive and altruistic behaviour, and many other qualities thought for centuries to be exclusively human.

Beauchamp (1999) claims that animals should have moral standing independent of their cognitive or moral capacity, based on two kinds of properties: having the capacity for pain and suffering, and properties of emotional deprivation. These are the properties that fuel the debates over research animals, animals in zoos and circuses and even farm animals. The injunction to avoid causing suffering, emotional deprivation, and many other forms of harm is as well established as any principle of morality... because harm is bad in itself... and not because the individual is or is not a moral person. Animals also have interests in not being deprived of freedom of movement and in continued life (Beauchamp, 1999).

Schinkel (2008) reviews Martha Nussbaum's writings on animal rights:

> Animals are entitled to opportunities to form attachments and to engage in characteristic forms of bonding and interrelationship. They are also entitled to relations with humans, when humans enter the picture, that are rewarding and reciprocal, rather than tyrannical. At the same time, they are entitled to live in a world public culture that respects them and treats them as dignified beings (Nussbaum, 2004).

The recent experience of a yearling black bear in Vancouver is an example of this entitlement. On December 13, 2011 conservation officers were called to the heart of Vancouver, British Columbia to tranquilize the bear which had been dumpster diving on the North Shore and was unknowingly picked up by a garbage truck which drove into town. It was tranquilized, prevented from falling to the road and then transported far from the city limits (Austin, 2011). The experience of Happy Feet the Emperor penguin is another example. He lost its way and ended up on Peka Peka beach on New Zealand's North Island. After being left alone for five days he was observed eating non-food items. Public pressure forced the Wellington

Zoo to accept gastroenterologist John Wyeth's offer to remove the debris from his gut and rehabilitate him for 72 days at the Zoo. An estimated 270,000 people followed his recovery over an internet livestream. Public pressure forced the National Institute of Water and Atmospheric Research to take him on board their vessel Tangaroa and release him 49 miles due north of Campbell Island. The public commitment to Happy Feet's welfare was seen in the complete scepticism shown to the scientists who attached a Sirtrack satellite tracker device and microchip to his body – the public felt that this would endanger him by attracting the attention of predators. It stopped working within days (Kennedy, 2011; Anon, 2011; Gonzoga, 2011).

The next proposition of Nussbaum (2004) neither Schinkel nor I agree with since it is against Nature:

> Animals are entitled to be able to live with concern for and in relation to animals, plants, and the world of nature'… This capability… calls for the gradual formation of an interdependent world in which all species will enjoy cooperative and mutually supportive relations with one another… it calls, in a very general way, for the gradual supplanting of the natural by the just.

Next Nussbaum makes a similar argument to that of Stone (1974, p.11). Stone argues that nature should count jurally – to have a legally recognised worth and dignity in its own right, and not merely to serve as a means to benefit "us." He claims that for a thing to be a holder of legal rights an authoritative body must review the actions and processes of those who threaten it and three additional criteria should be satisfied. The thing can institute legal actions at its behest; second, that in determining the granting of legal relief, the court must take injury to it into account; and; third, that relief must run to the benefit of it. However, nature with legal standing would still be dependent on the cultural values, wisdom and competence of the lawyer(s) chosen to represent it and those of the court of law with jurisdiction. Clayton (2000) claims that justice becomes more relevant in circumstances in which a desired response is scarce and in which there are citizens who ascribe moral significance and values to the environment, Nussbaum writes:

> For nonhuman animals, the important thing is being part of a political conception that is framed so

as to respect them and that is committed to treating them justly...On the material side, for nonhuman animals, the analogue to property rights is respect for the territorial integrity of their habitats, whether domestic or in the wild.

Nussbaum's property rights suggestion was tested in Rexburg, Iowa and in Trinidad. In the Iowa case a house was built on a winter snake sanctuary – (a hibernaculum) - where non-poisonous garter snakes hibernated. It was sold repeatedly to unsuspecting owners who could not cope (Alpert, 2011). The ethically-sound solution would be to turn the house into a Herpetology research station. In the case of Quesnel's neighbour he had already killed two small mapapire balsin snakes (40 cm) in his hedge before he was bitten by a third (Quesnel and Abdullah, 2009). He did not think they had a right to live that close to him. Within the fortnight, another neighbour of Quesnel's killed a mapepire balsin that was 115 cm long and another 42 cm mapapire balsin was killed on Quesnel's property by a landscaping machine. Quesnel postulated that all of the snakes were the offspring of one (now dead) female living close to the house of the neighbour with the hedge and that they had been born the September before.

The Criminal Code of Canada prohibits anyone from wilfully causing animals to suffer from neglect, pain or injury. This Code is enforced by police, the Prevention of Cruelty to Animals and/or provincial and territorial Ministries of Agriculture. Other Acts (Health of Animals Act, Meat Inspections Act), enforce humane transport and slaughter. Animal welfare seeks to improve the living, transport and slaughter conditions for farm animals and other animals being used by humans and is a more fleshed-out concept in organic farming (the widely used standards approved by the International Federation of Organic Agricultural Movements IFOAM) than in conventional agriculture (Lund, 2006; Jones, 2000; Buttel, 2000).

Organic farming systems are directed towards enhancing natural life cycles rather than suppressing nature.

In British Columbia organically-raised rabbits are given fresh cut sticks from any of the following trees to keep their teeth short: apple, willow, big-leaf maple, arbutus or red alder. In conventional rabbit production rabbits' teeth are cut with pliers. Commercial rabbit pellets typically contain four plant products (*Medicago sativa* L., *Hordeum vulgare* L., *Triticum aestivum* L. and *Glycine max* (L.) Merr.). Some pellets are made from *Phleum pratense* L., rather than *Medicago sativa* L.). Organic rabbit producers in British Columbia fed their rabbits at least twenty plants (Lans and Turner, 2010).

The EU legislation for farm animal welfare includes aspects of the rights-oriented approach and hinges on 5 freedoms: Freedom from discomfort; Freedom from hunger and thirst; Freedom from fear and distress; Freedom from pain, injury and disease; Freedom to express natural behavior - http://ec.europa.eu/food/animal/welfare/actionplan/actionplan_en.htm

The natural living approach is appropriate for snakes and caimans- expressing natural behavior and living a natural life (Lund, 2006). There are exceptions -a 15 metre long python, weighing 446 kg with a diameter of 85 centimeters that eats five dogs a month, lives in a recreation park in the central Java town of Kendal (Madigan, 2004) and obviously belongs there if there is no large-enough natural habitat for it to live in safe proximity to humans and their pets.

I cannot imagine that any caiman in Trinidad would live the life of Elvis the Australian crocodile. Estimated to be about 50 years old Elvis has been living as an attraction in the Australian reptile park in Gosford since 2008. In his latest escapade, on 28 December 2011 he lunged at the two men mowing the grass in his enclosure and dragged one of the mowers into the water with him where he guarded it for two hours, losing two teeth in the process. He was lured away with some kangaroo meat so that it could be retrieved. Apparently he was angered that two lawnmowers were used instead of one to save time. At a previous location (a breeding facility) he killed two female companions; however that 'domestic violence' story is also attributed to the previous inhabitant of the crocodile enclosure -Eric who lived at the reptile park from 1989 to 2007 after being implicated in the deaths of two children. Elvis ended up in captivity because he was boarding fishing boats in Darwin harbor (Jackson, 2011). Interestingly the fate of these two crocodiles violates one of Nussbaum's (2006) principles…"when there is a plausible reason for the killing (preventing harm to crops or people or other animals)… no entitlement based on justice has been violated."

Nor would the behavior of the late Crocodile Dundee Steve Irwin be tolerated in Trinidad, although animal cruelty exists and illegal cock fighting still takes place. Germaine Greer criticized Steve Irwin's treatment of animals in 2006. She wrote that the only animals he could not dominate were parrots and that he never hesitated to barge into animal's natural habitats to manhandle them. Greer had12 front-fanged venomous snake species in her Queensland rainforest and claimed that snakes were not inclined to unprovoked aggression and would get out of the way if given a choice. Irwin was apparently criticized only once in Australia – when he took his month-old son and a dead chicken into an Australian zoo crocodile enclosure (Greer, 2006).

Deckha (2008) makes a distinction between animal welfare initiatives and rights-oriented approaches, then devotes the rest of her paper to a discussion of how PETA uses sexual objectified women's bodies to capture attention and how PETA make references to the way marginalized human groups were treated in the past. Deckha writes that these campaigns could be seen as productive and subversive and not necessarily sexist and racist. This echoes the view I expressed in an Animal Science class as an undergraduate that publicity-seeking campaign are necessary because some farmers have to be forced to change (Figure 6).

Animal rights campaigners film 'shocking' conditions at rabbit farms. Monday 19 December 2011

Politicians and animal welfare campaigners have reacted with shock to film recorded secretly at eight of the Netherlands' rabbit farms, where animals are bred for consumption. The recordings, made by a group calling itself Ongehoord (unheard), shows rabbits with missing ears and other injuries living in cramped cages with wire floors. In some sequences dead animals are in the same space as the living. Animal health professor Frank van Eerdenburg told the AD that 'sick animals should not be with healthy ones.'

Regulations

The organisation also filmed the rabbit cages owned by Sjef Lavrijsen, who chairs the rabbit farmer association. 'It is really awful. No-one wants to see this,' he told the paper. 'No farmer wants dead or sick animals.' Lavrijsen said his farm meets new standards on rabbit welfare introduced by the government this year. By 2016, all 91 rabbit farms will have to meet new requirements, including plastic matting on the floor and giving rabbits something to gnaw on, the AD said.
http://www.dutchnews.nl/news/archives/2011/12/politicans_shocked_by_rabbit_f.php

The only rabbit farmer in the Netherlands with a 'Beter Leven' ranking (animal welfare label, initiated by the Dutch Society for the Protection of Animals), keeps most of his meat rabbits in the same conditions as on regular farms: crammed, ill, bored, and extremely stressed. Shockingly, the rabbits in Dutch farms are not protected by any Dutch law – the farmers play by their own rules. In 2006, Dutch rabbit farmers drafted animal welfare regulations for the commercial rabbit farming industry. The most important changes consist of enlarging the cages, adding cage enrichment and installing a platform for mother rabbits. In 2011, half of these regulations were put into practise. The current published investigation shows welfare improvements mean little to nothing for the animals.

http://achtergeslotendeuren.org/rabbit

Figure 6. Animal rights campaigners film 'shocking' conditions at rabbit farms

Conclusion

The environmental sceptic Peter Huber argues that humans have no moral obligation to non-human nature because humanity has the ability to dominate and control nature (Judeo-Christian doctrine) (Jacques, 2006). Deep anthropocentrism - enlightened anthropocentrism favours saving coral reefs for future medical benefits or for biodiversity reasons and for resources that humans use currently and that may be used for future generations, not for their own sake.

Siurua (2006) discusses how David Schmidtz divides morality into two parts: a "morality of personal aspiration," encompassing the ethical convictions and ideals according to which a person orients his or her actions, and a "morality of interpersonal constraint," which forms the basis of institutional arrangements to regulate social interactions between individuals pursuing their personal goals. Schmidtz argues that strict preservationism (in the sense of a rejection of any instrumental utilization of nature in protected areas, often motivated by nonanthropocentrism) may be acceptable and justified as part of a morality of the first kind, but as long as the costs of actual preservation are to be borne by people who do not share preservationist values, the promotion of preservationism as the foundation of interpersonal morality is doomed to failure and consequently ought not to be undertaken.

McCauley (2006) claims the push for ecosystems services to be quantified in order for Nature to be valued is akin to saying that civil-rights advocates would have been more effective if they provided economic justifications for racial integration. Nature conservation should be framed as a moral issue and argued as such to policy-makers, says McCauley, since policy makers are just as accustomed to making decisions based on morality as on finances.

Acknowledgements

The Centre for Gender and Development Studies of the University of the West Indies obtained funding from CIDA for the initial research which was published in 2000 as Governance aspects of the Nariva Swamp Nariva Swamp Case Study.

Nariva swamp Ecotourism project. Implemented in conjunction with the Caribbean Network for Integrated Rural Development (CNIRD), St. Augustine. Funded by AMOCO.

References

Aitken, Thomas, H.G. (1973). Bush Bush forest and the Nariva Swamp. *Journal of the Trinidad Field Naturalists' Club*, www.wow.net/.../Papers/BushBush/bushbush.html

Alpertt, Lukas I. (2011). Snake house of horrors: Idaho family driven from home after finding thousands of serpents. *NY DailyNews.com*, Wednesday, June 15, 2011. http://articles.nydailynews.com/2011-06-15/news/29681683_1_snake-problem-idaho-family-musk

Anon, (2009). It's not TT snake. *Newsday* Thursday, July 16 2009. http://www.newsday.co.tt/features/0,103861.html

Anon, (2010). ODPM [Office of Disaster Preparedness and Management] Ready for Evacuation. *Newsday*, Wednesday, August 11 2010.

Anonymous, (2010). Big stink in Aripero, *Newsday*, Friday, July 16 2010.

Anon, (2011). Did 'Happy Feet' Penguin Become Happy Meal? A satellite tracking device on the wayward penguin last sent a transmission on Sept. 4. http://news.discovery.com/animals/happy-feet-penguin-lost-110912.html

Asson, Cecily.(2010). Saved From Flood. *Newsday*, Wednesday, August 11 2010

Asson, Cecily. (2010). Snakes, caimans scare villagers. *Newsday*, Thursday, December 30 2010.

Austin, Ian. (2011). Black bear hitched a ride in garbage truck to downtown Vancouver. http://www.theprovince.com/news/Black+bear+hitched+ride+garbage+truck+downtown+Vancouver/5848762/story.html#ixzz1iAyferk0

Bacon, P. R., Kenny, J. S., Alkins, M.E., Mootoosingh, S.N., Ramcharan, E. K., and Seebaran, G.B.S.

(1979). Studies on the biological resources of Nariva swamp, Trinidad. *Occasional Papers No. 4 of the Zoology Dept.*, University of the West Indies, Trinidad.

Bacon, P.R. (1988). Freshwater foodchains of Caribbean island wetlands. *Acta Cientifica,* 2 (2-3), 74 – 93.

Bacon, Peter. (1997). Water management for sustainable development in the Nariva Swamp. In: CFCA, 1997. Nariva swamp seminar. Seminar held at Faculty of Agriculture, U.W.I., St. Augustine, Trinidad and Tobago. CFCA, Port of Spain, Trinidad

Bacon, P.R., Mahadeo, V.A. (1999). Impacts of agriculture on wetlands in Trinidad. Paper presented at Agriculture in the Caribbean: Issues and Challenges, UWI Ag.50.

Bailey, Cathryn. (2005). On the backs of animals: The valorization of reason in contemporary animal ethics. *Ethics and the Environment,* 10 (1), 1 – 17.

Barrio-Amorós, C. L. and R. Manrique. (2007). Observations of natural history of the green Anaconda (*Eunectes murinus* Linnaeus, 1758) in the Venezuelan Llanos. An ecotourist perspective. Fundación AndígenA. 34 pages.

Beauchamp, Tom L. (1999). The failure of theories of personhood. *Kennedy Institute of Ethics,* Journal 9 (4), 309-324.

Begley, Sharon. (2006). Animals seem to have an inherent sense of fairness and justice. *Wall Street Journal* 10 November 2006 p. B 1.

Boos, Hans E.A.(2001). *The snakes of Trinidad and Tobago*. Texas A&M University Press, College Station, TX.

Bonadie, Wayne A., Bacon Peter, R. (2000). Year-round utilisation of fragmented palm swamp forest by red-bellied macaws (*Ara manilata*) and orange-winged parrots (*Amazona amazonica*) in the Nariva Swamp (Trinidad). *Biological Conservation,* 95, 1-5.

Bourdieu, P. (1977). *Outline of a Theory of Practice*. New York: Cambridge University Press.

Brosnan SF, De Waal FB.(2003). Monkeys reject unequal pay. *Nature,* 425(6955), 297-9.

Brydon, Anne. (2006). The predicament of Nature: Keiko the whale and the cultural politics of whaling in Iceland. *Anthropological Quarterly,* 79 (2), 225-260.

Buttel, Frederick (2000). The recombinant BGH controversy in the United States: Toward a new consumption politics of food? *Agriculture and Human Values,* 17 (1), 5–20.

Castree, Noel. (1999). Contested natures. Review. *Transactions of the Institute of British Geographers. New Series,* 24 (4), 510 – 511.

Catton, W., Dunlap, R. (1978). Environmental Sociology: A new paradigm. *The American Sociologist*, 13, 41–9.

CFCA, (1997). *Nariva swamp seminar.* Seminar held at Faculty of Agriculture, U.W.I., St. Augustine, Trinidad and Tobago. CFCA, Port of Spain, Trinidad.

Clapp. R.A (2004). Wilderness ethics and political ecology: Remapping the Great Bear Rainforest. *Political Geography,* 23, 839–862.

Clayton, Susan. (2000). Models of justice in the environmental debate. *Journal of Social Issues,* 56 (3), 459 – 474.

Cornwall, A. (1998). Gender, participation and the politics of difference. In: Irene Guijt and Meera Kaul Shah (eds), *The Myth of Community: Gender in Participatory Development*, IT Publications.

Cross, N., Dator, J., Durbal, S., Wahab, A. (1999). A pilot study of the gendered socio-cultural, socio-economic, and governance issues in the Kernahan-Cascadoux community, Nariva. Project document – Building Approaches Towards Sustainable Livelihoods. CGDS, CIDA, ISLE.

Deckha, Maneesha, (2008). Disturbing images: PETA and the feminist ethics of animal advocacy. *Ethics and the Environment,* 13 (2),35-76.

Doodhai, M. (2000). A simple life. Kernaham is one of South's hidden treasures. *Sunday Guardian* June 4, 2000 (pp. 8).

Draft Second Five - Year Plan 1964 - 1968. National Planning Commission: Dr. Eric Williams, Mr. Arthur, N. Robinson, Mr. John O'Halloran, Mr. Robert Wallace, Mr. Lionel Robinson, Mr. Louis Alan Reece, Mr. Jack Harewood, Mrs. Patricia Robinson, Mr. David Weintraub, Mr. William Demas. Government Printery, Trinidad and Tobago, 1963. Chapter XIV (pp. 173, 179, 201-203).

Duguid, A., Downie, R., Heath, M., Hambly, C. (1996). A comparison between fish communities in Nariva Swamp and adjacent rice paddies. Report of the University of Glasgow Exploration Society, (pp 22 – 27).

Driver, T. (2001). Watershed management, land tenure and forests in the Northern Range, Trinidad. In:

Leach Melissa, Amanor Kojo and Fairhead James. 2001. Forest science and forest policy: Knowledge, instituions and policy processes. Final Report to ESCOR of the Department for International Development (DFID), Project No. R7211 http://www.ids.ac.uk/ids/KNOTS/PDFs/ForestESCORReport.pdf.

Dunlap, Riley E., Catton, William Jr. (1979). Environmental sociology. *Annual Review of Sociology,* 5, 243-73.

Durbal, Sharda. (2000). Natural resource use and management in Kernahan and Cascadoux. Nariva Swamp: A gendered case study: Social, cultural and gendered analysis of Kernahan and Cascadoux. The Centre for Gender and Development Studies, The University of the West Indies, St. Augustine.

FAO, (1957). Report to the government of Trinidad and Tobago on the reclamation of the Caroni, Oropouche and Nariva areas. Report No. 636. Food and Agriculture Organisation, Rome. 111 pp.

Foucault, Michel. (1980). *Power/Knowledge: Selected Interviews and Other Writings, 1972–1977,* Colin Gordon (Ed.). New York: Pantheon.

Gonzaga, Shireen. (2011). New Zealanders help a wayward penguin find his home. Earth Sky blog. http://earthsky.org/biodiversity/new-zealanders-help-a-wayward-penguin-find-his-way-home

Greer, Germaine. (2006). 'That sort of self-delusion is what it takes to be a real Aussie larrikin'. http://www.guardian.co.uk/world/2006/sep/05/australia

Hardjoeno, Vanderzwaag, D., Shaw, T., Bedeno, J.A.S., Sirju-Charran, G. (1996). Governance. Position paper. Workshop on core-course of Island Sustainability, Livelihood and Equity. Ujung Padang, 1 - 6 December 1996. Indonesia.

Hosein, G., Cross, Nicole. (2000). Nariva Swamp: A gendered case study: Social, cultural and gendered analysis of Kernahan and Cascadoux. The Centre for Gender and Development Studies, UWI, St. Augustine.

Huggan, Graham. (2004). "Greening" Postcolonialism: Ecocritical Perspectives. *MFS Modern Fiction Studies,* 50(3), 701-733.

Institute of Marine Affairs, (1998). Final Report for the Environmental Impact Assessment of the Nariva Swamp (Biche Bois Neuf Area). IMA/MALMR, Trinidad and Tobago.

Institute of Marine Affairs, (1999). Final Report: Formulation of the Nariva Swamp Management Plan. IMA/MALMR, Trinidad and Tobago.

Jackson, C. (1993). Environmentalisms and gender interests in the Third World. *Development and Change,* 24, 649 - 677.

Jackson, Joe. (2011). All Shook Up: Elvis the Australian crocodile is tired of your yard work. http://newsfeed.time.com/2011/12/29/all-shook-up-elvis-the-australian-crocodile-is-tired-of-your-yard-work/

Jacques, Peter. (2006). The rearguard of modernity: Environmental skepticism as a struggle of citizenship. *Global Environmental Politics,* 6,76 – 101.

James, K. (1994). *The Potential of Bush bush wildlife sanctuary, Trinidad for ecotourism.* Undergraduate thesis, B.Sc., Zoology/Chemistry. UWI. 97 pp.

Jones, Kevin.(2000). Constructing rBST in Canada: Biotechnology, Instability and the Management of Nature. *Canadian Journal of Sociology,* 25 (3), 311-341.

Kacal, S., Homer, F. (1993). Recommendations for 3 National Parks; Matura, Nariva and North East Tobago, unpublished.

Kacal, S. (1997). Nariva issues: looking toward community co-management. In: Nariva swamp seminar. Seminar held at Faculty of Agriculture, U.W.I., St. Augustine, Trinidad and Tobago. CFCA, Port of Spain, Trinidad (pp. 60.)

Kacal, S. (2000). Social assessment and community action plan of Nariva managed resource are. Final report prepared for Ministry of Agriculture, Land and Marine Resources, Government of the Republic of Trinidad and Tobago and World Bank. Unpublished.

Kalema-Zikusoka, Gladys.(2005). Protected areas, human livelihoods and healthy animals: Ideas for improvements in conservation and development interventions. In: Osofsky, Steven A., Cleaveland, Sarah, Karesh, William B., Kock, Michael D., Nyhus, Philip J., Starr, Lisa, Yang, Angela, (eds.). Conservation and Development Interventions at the Wildlife/Livestock Interface. The World Conservation Union (IUCN).

Keeler, A.G.and Pemberton, C. (1996). Nariva swamp: An exercise in environmental economics. Notes from a seminar presented at the University of the West Indies, St. Augustine, Nov. 20, 1996.

Kennedy Maev, (2011). Happy Feet the penguin's tracker falls silent. Guardian.co.uk Mon 12 Sept 2011.
http://www.guardian.co.uk/environment/2011/sep/12/happy-feet-penguin-tracker-silent

Kupfer, J.H. (2003). Engaging Nature aesthetically. Journal of Aesthetic Education 37 (1), 77 – 89.

Lans, C. (1996). The price of rice. *Express*. Monday July 15, 1996. pp. 15.

Lans, C. (1999). Nariva swamp Ecotourism project. Implemented in conjunction with the Caribbean Network for Integrated Rural Development (CNIRD), St. Augustine, Trinidad and Tobago.

Lans C. (2001). Governance aspects of the Nariva Swamp Case Study. Centre for Gender and Development Studies, University of the West Indies (UWI), St. Augustine, Trinidad and Tobago.

Lans C.(2007). *Politically Incorrect and Bourgeois: Nariva Swamp Is Sufficient Onto Itself*. Self-published book originally available at Lulu.com, now on Amazon's CreateSpace.

Lans C, Turner N. (2010). Rearing and Eating Locally-Grown Organic Small-Scale Poultry and Rabbits in British Columbia, Canada. *Current Nutrition & Food Science,* 6(4), 290-302.

Leach, Melissa, Fairhead James (2001). Science, policy and national parks in Trinidad and Tobago. Working Paper from the project 'Forest Science and Forest Policy: Knowledge, Institutions and Policy Processes'. Also presented at the Workshop 'Changing perspectives on forests: ecology, people and science/policy processes in West Africa and the Caribbean', 26-27 March 2001 at The Institute of Development Studies, University of Sussex.

Lee, M. and Jacque, A. (1993). The rice sub-sector. Agricultural planning division working paper no. 3. (revised).

Lund, Dr. Vonne (2006) Natural living – a precondition for animal welfare in organic farming. *Livestock Science*, 100 (2-3), pp. 71-83.

Madigan, Charles (2004). Snakes alive, it's the tail end of 2003. *Chicago Tribune* Jan 04, 2004.
http://articles.chicagotribune.com/2004-01-04/news/0401040199_1_mad-cow-disease-reticulated-python-snake

Maharaj, K. (2000). A simple life. *Sunday Express* Section 2, May 14 2000 (pp. 31).

Mahy, M. (1997). *Feasibility of co-managing a wetland of international importance: the case of the Nariva Swamp, Trinidad*. Unpublished thesis. Master of Environmental Studies, Dalhousie University, Halifax, Nova Scotia, Canada.

MALMR, (1983). A draft national policy for food and agriculture. Republic of Trinidad and Tobago. May 1993. Attachment 1.

MALMR, (1995). Review of the agriculture sector 1993. Prepared by the agricultural planning division, Ministry of Agriculture, Land and Marine Resources. Jan 1995.

McCauley, D.J. (2006). Selling out on nature. *Nature,* 443 (7107), 27-8.

Miller, Daniel (1994). *Modernity, An Ethnographic Approach: Dualism and Mass Consumption in Trinidad*, Oxford and Providence: Berg.

Ministry of Agriculture, Lands and Marine Resources, 1994. Draft. White Paper. Food and Agriculture Policy.

Mootoosingh, S.N. (1979). The growth of conservation awareness in Trinidad and Tobago (1965-1979). Occasional paper no.3, Dept. of Zoology, UWI (St. Augustine).

Murdoch, J. (2001). Ecologising sociology: Actor-Network Theory, co-construction and the problem of human exemptionalism. *Sociology*, 35(1), 111-133.

Murphy, Raymond (1995). Sociology as if Nature did not matter: An ecological critique. *The British Journal of Sociology*, 46 (4), 688 – 707.

Murphy, John, C. (2011). Trinidad's Coastal Swamps and Snakes. http://herpetologytt.blogspot.com/2011/08/trinidads-coastal-swamps-and-snakes.html

Nathai-Gyan, Nadra (1997). Conservation status of the Nariva wetland. In: CFCA, 1997. Nariva swamp seminar. Seminar held at Faculty of Agriculture, U.W.I., St. Augustine, Trinidad and Tobago. CFCA, Port of Spain, Trinidad.

National Policy on Wetland Conservation. Trinidad and Tobago. Draft prepared by the National Wetland Committee. January 16, 1996.

NEDECO, (1981). A summary and review of previous studies of the Nariva Swamp. Technical Note No. 4. Netherlands Engineering Consultants. 40 pp.

NEDECO, (1983). Phase 1 - Investigations on the development of the Nariva Swamp. Vol. 1. Main Report. Final Report. Ministry of Agriculture, Lands and Fisheries, Port of Spain. 36 pp.

Nussbaum, Martha C. (2004). Beyond "Compassion and Humanity". Justice for Nonhuman Animals. In Sunstein, Cass R. / Nussbaum, Martha C. (Ed.), Animal rights: current debates and new directions

New York: Oxford Univ. Press (pp. 299-320).

Nussbaum, Martha C. (2006). *Frontiers of Justice*: Disability, Nationality, Species Membership. Cambridge, MA: Harvard University Press, Belknap Press.

OCTA, (1970). Nariva Swamp Development Project. Feasibility Report. Overseas Technical Co-operation Agency of Japan. 212 pp.

Paul, F. (1998). Farmers call for IMA report on Nariva Swamp. *Trinidad Guardian* Friday Oct 16, 1998 (pp. 5).

Paul, Anna-Lisa (2011). Met Office: Take All Precautions. *Newsday* Monday, October 17 2011.

Quesnel, Victor C., Abdullah Gail (2009). Mapepires Galore. The Field Naturalist Issue no. 3/2009. July-September 2009.

Ramsar Convention (1995). Monitoring Procedure, Nariva Swamp, Trinidad and Tobago. Gland, Switzerland.

Ramsar Convention (1996). Final Report, Monitoring Procedure, Nariva Swamp, Trinidad and Tobago. Ramsar Convention. Gland, Switzerland.

Review of the Agriculture sector 1993 (1995). Prepared by the Agricultural Planning Division, Ministry of Agriculture, Land and Marine Resources. T&T.

Rostant, R. (1998). Nariva comes back from the brink – Rice farmers want to be allowed back. *Trinidad Guardian* February 2, 1998.

Sasa M, Wasko DK, Lamar WW. (2009). Natural history of the terciopelo *Bothrops asper* (Serpentes: Viperidae) in Costa Rica. *Toxicon,* 54(7),904-22.

Schinkel, Anders. 2008. Martha Nussbaum on Animal Rights. *Ethics & the Environment,* 13 (1), 41-69.

Seereeram, A., Luke, E., Seales, J., Ramdeen, P., Indalsingh, T. (1991). Final report 91.04.03. Integrated rice mechanisation project proposal. Technical assistance project between the Government of the Republic of Trinidad and Tobago and the Japanese Government. Ministry of Food Production and Marine Exploitation.

Siurua, Hanna (2006.) Nature above people: Rolston and "Fortress" conservation in the South. *Ethics & the Environment,* 11 (1),71-96.

Sletto Bjørn (2002). Producing space(s), representing landscapes: Maps and resource conflicts in Trinidad. *Cultural Geographies,* 9, 389–420.

Stone, Christopher, D. (1974). *Should trees have legal standing? Toward legal rights for natural objects.* California: William Kaufman, Inc.

Taitt, G. (1999). Rice, culture and government in Trinidad 1897 – 1939. In *Colonial Caribbean in Transition: Essays on Postemancipation Social and Cultural History.* Bridget Brereton and Kelvin A. Yelvington (Eds.). Gainesville: University Press of Florida; Kingston: The Press, UWI.

Taylor, K., H. P. Nelson, and A. Lawrence. 2011. Population Density of the Cook's tree Boa (Corallus ruschenbergerii) in the Caroni Swamp, Trinidad. In A. Lawrence and H. P. Nelson (Eds.).Proceedings of the 1st Research Symposium on Biodiversity in Trinidad and Tobago (pp. 8 – 18), University of the West Indies Department of Life Sciences.

Thelen, K.D. and Faizool, S. (1980). Plan for a system of national parks and other protected areas in Trinidad and Tobago. Forestry Division, Ministry of Agriculture Lands and Fisheries, POS, Trinidad. 106 pp.

Tompkins, E., Adger, W .Neil (2002). Institutional networks for inclusive coastal management in Trinidad and Tobago. *Environment and Planning A,* 34,1095 – 1111.

T&T Feb08 TRep (John). www.naturalist.co.uk/reports2008/trinidad208.pdf

Wahab, A. (1997). The status of women in the Nariva wetland communities: A case study. Unpublished.

Wahab, A.S. (1997). Stakeholders' perceptions of natural resource conflict - the case of the Nariva wetland, Trinidad and Tobago. Unpublished M.Sc. Thesis, Shimane University, Japan.

White paper on Agriculture (1979). Ministry of Agriculture, Lands and Fisheries. Government printery, Trinidad and Tobago.

Wildlife Section (1993). Historical perspectives on habitat destruction in the Nariva Swamp, Trinidad: A Wildlife Section Issues Paper. Wildlife Section, Forestry Division, Ministry of Agriculture, Land and Marine Resources, St. Joseph, Trinidad and Tobago.

Willems-Braun, Bruce (1997). Buried epistemologies: The politics of Nature in (post) colonial British Columbia. *Annals of the Association of American Geographers,* 87 (1), 3-31.

Wolfe, Cary (2003). Old orders for new: Ecology, animal rights, and the poverty of humanism. In: *Animal Rites: American Culture, the Discourse of Species, and Posthumanist Theory.* Chicago: U of Chicago. Pp. 21-43.

Worth, C. Brooke (1967). *A naturalist in Trinidad.* Philadelphia and New York: J.B. Lippincott

Appendix: List of acronyms

SEAGA	Socioeconomic and Gender Analysis (FAO)
FAO	Food and Agriculture Organisation
EMA	Environmental Management Authority
CFCA	Caribbean Forest Conservation Association
IMA	Institute of Marine Affairs
ISLE	Island Sustainability, Livelihood and Equity Programme
MALMR	Ministry of Agriculture, Land and Marine Resources
UWI	University of the West Indies
CAREC	Caribbean Epidemiology Centre
GORTT	Government of the Republic of Trinidad and Tobago
PRA	Participatory Rural Appraisal
NEDECO	Netherlands Engineering Consultants
TRVL	Trinidad Regional Virus Lab
O.C.T.A	Overseas Technical Co-operation Agency
EIA	Environmental Impact Assessment
CIAT	Centre Internacional de Agricultura Tropical
JICA	Japanese International Corporation Agency
NFM	National Flour Mills
GEF	Global Environment Facility
WLCC	National Wetlands Committee
CITES	Convention on International Trade in Endangered Species of Wild Fauna and Flora

Eco-alternatives to slash and burn agriculture for T&T

Abstract

Sustainable agriculture is the best alternative for Trinidad and Tobago which has the problems of increasing soil erosion and degradation, and shortages of livestock fodder. Historically land was given out for farming without any consideration of their suitability for the purpose this has led to soil loss and flooding causing property damage. Hillside farmers could be encouraged to develop mixed farming systems including livestock production which would facilitate the incorporation of grasses and fodder trees for livestock feed. Labor intensive traditional agriculture with its emphasis on multiple cropping, multiple storeys and multiple uses are more ecologically sound in the humid tropics than conventional agriculture.

Keywords: agroforestry; Trinidad and Tobago; soil erosion; flooding

Highlights

- Agroforestry, cover cropping or multiple cropping are ways to manage heavy tropical clay soils on steep slopes in the Caribbean.

- Agroecosystems should imitate the structure of the primary forest or the secondary regrowth.

- Crop associations and multiple cropping can facilitate productive and continuous land use.

1.   Introduction

This paper reviews the literature on soil types and agricultural practices and makes the case that agroforestry, multiple cropping and low-input farming can reduce flooding locally and provide effective methods of soil and water management for crop and livestock production. The most sustainable systems would be self-providing in the following resources:

*Mineral nutrients from the deeper soil layers

*Nitrogen fixation

*Immobilized phosphorus reserves in the soil

*Manure, compost, or mulch integrated into the system

*Constant or expanding pool of organic nutrients

Nutrient losses would be reduced by minimizing leaching, erosion, fixation of elements and increasing waste recycling.

Sustainable and low input land use systems such as multiple cropping and agroforestry have multiple outputs and maintain or enhance the production base. They are also socially acceptable since traditional home gardens follow a similar multiple storey-multiple-use system. There is also a historical basis for agroforestry since agro-silviculture flourished for many years in Trinidad and Tobago as part of the establishment of mixed teak and cocoa plantations. Plantation owners leased plots of land for several years to workers who cleared the forest, planted cocoa and used the area for their own food crops until the cocoa was established.

1.1. Trinidad

In Trinidad squatters with no incentives for appropriate soil management often occupy the steep slopes in the Northern Range. The system of land tenure from colonial times has been freehold and this, according to (Beard, 1946), has permitted the widespread abuse and neglect of land. Shifting cultivation is practiced by both squatters and by small farmers with legal title creating a serious problem of water control and soil erosion in the wet season. Burning the forest cover removes weeds and mineralizes the available nutrients that accumulated in the wood, the resulting ash provides a rich seed bed. The farmer then plants a range of crops depending on home needs and market requirements. Row crops, usually corn *Zea mays* and pigeon peas *Cajanus cajan*, are grown without any soil and water conservation practices. Trinidad's clay soils become more acid when they are drained (Gumbs, 1982). So when productivity declines the plot is abandoned and the cycle started anew elsewhere. The vegetation that succeeds in these former clearings is essentially scrub and poor quality grasses, with a few remaining individuals of the former crop species. The grasses have an extensive root system that is not destroyed by fire and help them compete successfully against other species. Recurrent slashing and burning of the woody shoots leads to a depletion of their root starch reserves and prevents successional recolonization.

Farmers prefer to use the slopes in the wet season when the low lying areas are flooded. Cooler temperatures in the hills facilitate the growing of flowers, herbs and vegetables. Because of the limited access to the plots, crops of small bulk and high value are chosen. Little fertilizer is used and harvested produce must be manually transported due to bulkiness and also mechanical damage (Ahmad, 1991). However hillside cultivation adds to the other factors that create flooding in the plains; the inadequacy of the existing drainage system, the over-flowing of river banks in the wet season, and the large volume of uncontrolled runoff from the hills which besides the squatters is also partially caused by inappropriate construction activities (Gumbs, 1982). These problems are aggravated by the low relief and poor drainage of the low lying areas especially in the northern plain. Heavy soil types add to the slow removal of water and the consequent prolonged flooding in some areas. Various projects to minimize landslides have been documented in Anderson et al., (2011). A table of all flooding taking place in Trinidad from 1981 to 2006 is available online at www.cso.gov.tt/files/statistics/Chapter%2012.pdf (GORTT, 2007).

Excerpts from recent newspaper reports show the impact of flooding:

6. A villager from Debe (on the West side of Trinidad with Aquentic Chromuderts) complained about poor drainage leading to an influx of reptiles (Asson, 2010).

7. While Met Office officials are hopeful the Caroni River [originating in the Northern Range] will not burst its banks, they have advised residents in low lying areas to take all possible precautions (Paul, 2011).

8. The Cipero River broke its banks and ..set off massive flooding (Asson, 2010).

9. Residents at Roberts Street and Emmanuel Trace complained yesterday that former Caroni lands which were converted into rice fields were covered with water, and this resulted in several homes being flooded. (Anon, 2010).

10. Apart from the mounds of rubbish covering the land, residents said that the water-logged nature of the lot of land is a haven for scorpions, rats, caimans and snakes (Anonymous, 2010).

There is little legislation or enforcement of existing laws to control the development of the steep lands by private owners or squatters (Gumbs, 1982). The Government's main solution to the flooding and drainage requirements of the Caroni river basin watershed in the northern plains of Trinidad is national watershed management. The Northern Range Reforestation Project, a collaboration between the forestry sector which was responsible for reforesting approximately 16

000 ha. of land above the 225 m contour, and the agriculture sector which had the task of planting approximately 8,000 ha. of land between the 125 and 225 m contour with agricultural tree crops. The forestry sector planned to plant 250 ha. per year and the agriculture sector, 60 ha. per year. Demonstration sites showing basic conservation practices have also been established.

1.2. Tobago

In Tobago the hilly areas are left to the smaller farmers because the low lying areas have been historically divided into large estates owned by a few families. These large estates are mainly in extensive livestock production since the soil is based on tertiary rock and covered by coral limestone; 79 % of the soils are steepland soils. Some of the small farmers in Tobago have adopted contour planting and the use of an A-frame in the making of ridges. In Trinidad there seems to be less long term consideration of the land and furrows often follow the slope. A cultural consideration may be the presence in Trinidad of squatters from islands with different soil types. In the Windward islands (St. Vincent, St. Lucia, Dominica), where there are soils with andic properties, there was less soil erosion than in Trinidad. On these islands, bananas, sweet potatoes and edible aroids were being grown on 58 percent slopes with no great soil losses except for gullying (Ahmad, 1991).

There have been projects implemented in St. Lucia and Dominica to prevent landslides (Holcombe and Anderson, 2010). In St Lucia, Ahmad and Sheng proposed three classes of erodibility for steep soils as follows: stable soils---Andisols, Mollisols, some Inceptisols, and some Alfisols (Oxisols are included in this group if they are present); less stable soils --Ultisols, some Alfisols, some Inceptisols, and some shallow Mollisols-- fragile soils---Vertisols. Different levels of treatment, crops, cropping systems, and settlements can be prescribed for each group on comparable slopes if the soils are mapped in sufficient detail for this type of approach, and enforcement can be carried out.

2. **General background of Trinidad and Tobago**

2.1. Trinidad

The island of Trinidad lies 10 degrees 2 minutes and 11 degrees 21 minutes north latitude and between 60 degrees 30 minutes and 61 degrees 56 minutes west longitude. It is generally considered as the most southerly of the West Indian islands, but is more properly regarded as an outlier of the South American continent from which separated in recent times. It lies approximately 18 km. east of Venezuela. The island measures 104.6 km. in width and 80.5 km. in breadth and has an area of 4769 km squared. Trinidad belongs structurally to the South American continent and its geological history is quite distinct from that of the volcanic Antillean islands. Except for an intrusion of andesite on the north coast at Sans Souci, all rocks are sedimentary (Beard, 1946; Chenery, 1949).

2.2. Tobago

Tobago lies about 35 km. north-east of Trinidad, and has a length of about 40.2 km. from WSW to ENE (Beard, 1946). Its area is 330 km. squared. The main ridge forms the backbone of the island and is about 18 km. long. It keeps much nearer to the north coast than to the south. The general elevation of the ridge itself rises to a maximum of 576 m. The ridge contains 5 watersheds. There are numerous well-watered and fertile valleys. The western part of the island is hilly; elevations of over 153 m. are common and there are no flat areas of any extent except in the southwest corner. In this corner there is 10 square miles of flat land less than 31 m. in elevation.

2.3 Topography of Trinidad

Trinidad consists of three mountain ranges with flat or rolling land between the ranges. According to Gumbs (1982), Trinidad can be divided into nine topographical areas:

(1) Northern range

A mountainous belt, approximately 15 km. wide, with a precipitous northern boundary abutting onto a rocky coast.

(2) Northern terraces

These are sandy terraces occurring up to approximately the 100 m. contour in the eastern part of the island and lie level against the southern flanks of the northern range foothills.

(3)Northern plain

These terraces spread out in broad, level sheets, dissected by deep-cut river channels. They are now the flood plains of the Caroni and Caparo rivers overlain by recent alluvium.

(4) Northern swamps

The chief of these is the Caroni swamp on the west cost.

(5) Central range

This runs diagonally across the island and the highest point is approximately 350 m. above sea level.

(6) Naparima peneplain

All the undulating land south of the central range. It carries river systems flowing west, east and south, associated with swamps and coastal lagoons.

(7) Southern swamps

The chief is the Nariva swamp. The Oropuche and Roussillac swamps are much smaller wet areas on the southwest coast.

(8) Southern range

This is an irregular broad belt of low sandstone hills 65-175 m. high and running along the southern coast.

(9) Cedros peninsula

This forms the southwest extremity, some 55 km. long and comprises mostly undulating to hilly country.

2.4. Rainfall

In Trinidad there is a marked dry season extending from January to April/May followed by a wet season till year end which is often interrupted by a short dry spell, referred to as the *petit careme*, in October. Because of the direction of the tradewinds over Trinidad and the mountain ranges, the annual rainfall decreases from northeast to southwest from ~250 cm. to ~125 cm. At the highest elevations in the northeast, the annual rainfall may be as high as 300 cm. Often 80-90 per cent of the annual rain falls in the wet season. Sometimes 30-35 per cent of the annual rainfall occurs in June and July. An intense rainstorm can produce 8-10 cm. of rainfall and the rainfall intensity can be as high as 10 cm. per hour for short durations (Gumbs, 1982).

2.5. Geology and soils

Soil type and soil depth are important in designing agricultural systems that include soil conservation. The ease with which soil aggregates disintegrate on raindrop impact is a fairly good indicator of the erodibility of the soil. Igneous rocks and red soils on limestone develop well aggregated soils because the iron oxides help hold them together. Fragmentary volcanic materials with andic properties develop erosion-resistant soils (Weischet and Caviedes, 1993). More than half of Tobago is of igneous origin or epidiorite. This rock is known locally as 'rotten rock', since it decomposes easily creating deep rich and fertile soils with good physical properties. Soft parent material like shales, alluvial and colluvial deposits, or fragmentary volcanic materials, can be stabilized by tree and plant roots. Simple conservation techniques like maintenance of some forest cover; reduced tillage; grass barriers on the contour; contour drains, step drains, and grassed waterways may be enough depending on the slope (Ahmad, 1991).

The north and north east of Tobago have decomposed schists while one third of Tobago in the north and north-east have sedimentary rocks without limestone. The schists have formed good deep soils. If the parent rock is indurated, igneous and metamorphic rocks and limestones, there is a sharp contact between the hard rock and the soil. Soils formed on shales, phyllites, and sandstones are easily eroded. They have an abrupt boundary between the soils and the underlying rock. The rock can be layered and fine-grained so that tree roots have difficulty growing into and holding the soils. On slopes the water seeps through the soil and accumulates at the sharp rock/soil contact. The accumulated water then acts as a lubricant on which the overlying soil can slide (Ahmad, 1991). Land clearing and agriculture can aggravate this type of erosion, especially early in the wet season on bare soils which become supersaturated. Dislocations, crop losses and destruction of any anti-erosion structures result. Newly cleared land with the resulting distinct change in hydrological conditions, are prone to this type of erosion (Ahmad, 1991).

Larger stream flows in deforested sites transport more and larger particles downstream. On agricultural land higher rates of erosion can be continuous unless practices such as no-till agriculture or agroforestry are adopted. Land slips also occur in clay sediments. The parent materials of Vertisols have well-developed slip faces. These soils develop wide, deep cracks in the dry seasons. Cracking and desiccation becomes more severe on bare soils. Water from the first rains can easily accumulate before swelling and sealing of the cracks can occur. The result is massive land slips resulting in damage to agricultural roads, land and buildings (Ahmad, 1991).

Gullying is another form of erosion that occurs on steep land. This is common on porous volcanic soils and on Vertisols. Rapid supersaturation leads to the soil breaking off and gully formation. Removing the vegetative cover leads to drainage water from cultivated fields coursing down the slope, foot paths or field boundaries creating gullies (Ahmad, 1991). These erodible soils developed on hard parent rocks, need erosion prevention rather than control.

Trinidad's Northern range is Lower Cretaceous or older and consists of limited quantities of limestone and large areas of metamorphic schists. There are micaceous and talcose-quartzitic sandstones. On non-limestone areas the soil is poor. On the limestone areas near the hill tops the soil is a good rich friable fertile loam (Beard, 1946).

These rocks produce soil in which either the sand or the silt fraction is high, and there is not enough clay-size material and iron oxide to cement the soil. Rich muscovite deposits occur in the weak structured soils. The clay fractions and weak aggregates are easily broken up by raindrops. The mica flakes settle on their flat axes in the film of water on the soil surface leading to soil crusting (Weischet and Caviedes, 1993). Three soil series in the north are found in Arena, River Estate and Piarco. Arena has Orthoxic Quartzipsamments sand with 11% clay and 1% organic matter (Ekwue and Bartholomew, 2011). River Estate has Fluventic Eutropepts sandy loam with almost 17% clay and close to 1% organic matter. Piarco also has a sandy loam (Aquoxic Tropudults) with more silt and organic matter than the River Estate soil. Water runs off this crust increasing the disintegration of soil aggregates and transport of colloidal soil material. The water infiltration into a crusted soil may be less than one-fifteenth that in the same soil with a protective layer of surface mulch. The crust has a negative effect on seedling emergence, denitrification and toxic effects due to ethylene production. The lack of gaseous exchange leads to anaerobic conditions. A recent study on surface runoff found that Maracas clay loam (30.6 % clay content) had lower soil loss than Talparo clay (46.3 % clay content) or Piarco sandy loam (Ekwue and Harrilal, 2010). The study also found that an increase in soil slope increased the runoff in Talparo clay more than the other two soil types.

3. **Physical soil factors leading to long-term failure of short fallow subsistence agriculture.**

Many soils in the humid tropics are characterized by two-layered clay minerals with weak adsorption capacity. The inorganic mineral substances of the (oxisols, ultisols, xanthic and orthic ferrasols) are the result of a lengthy allitic weathering process and are rapidly exhausted. The clay minerals of the leached and desilified soils are usually 1:1 layered kaolinites (Weischet and Caviedes, 1993). These soils are poor in minerals, nutrients and cation exchange capacity (CEC).

Under an undisturbed natural forest, soil nutrients and CEC are bound to organic matter, i.e. humic acids. Because organic matter mineralizes at least five times faster in tropical than in temperate climates, the main carriers of nutrients and CEC disappear from the soil within 2 or 3 years after conversion from forest to biomass-exporting agriculture (Weischet and Caviedes, 1993). Talparo clay loam holds more water and has more cations than Maracas clay loam and Piarco sandy loams (Ekwue and Bartholomew, 2011).

Plant uptake and transpiration of soil water and mineral nutrient uptake are usually decreased for at least 2-3 years even in sites that rapidly regrow to forests. With reduced evapotranspiration, water flux through the soil is increased, and so are losses of nutrients through leaching to groundwater and streams. There are increases in soil temperature and moisture with a corresponding increase in rates of decomposition and nutrient mineralization in deforested sites. The forest floor decomposes rapidly, and, without regeneration, will eventually disappear. Rates of decomposition of 0.44% and 0.45% per day have been measured for cocoa leaves and Mora leaves respectively in Trinidad (Ahmad et al, 1982). High temperature and humidity accelerate the degradation process. The cation exchange capacity (CEC), or capacity of the soil to hold cations in a form available to plants, is generally lower than 15 meq/100g of soil, and the saturation (V) is lower than 30% (Weischet and Caviedes, 1993). Because poor soil structure results in a high erosion rate, weak water-retention capacity and a reduction of gaseous exchanges, mineral fertilisation with increased yields is not cost effective and may increase leaching. The low CEC of these soils reduces their response to modern agrotechnology especially because of the frequent and heavy downpours of rain.

4. Nutrient Cycling

In the humid tropics shifting cultivation may have evolved as a special method to deal with problematic soils. However it is unlikely that continuous and fertilizer-intensive cropping systems can replace the biomass-rich forest and maintain fertile and sustainable systems (Weischet and Caviedes, 1993).

The net primary production (NPP) of any ecosystem is dependent on the availability of resources, and the advantage plants can make of them. Converting natural ecosystems to agricultural uses changes both the nutrient availability and the way that plants use resources. Ecosystems with a positive NEP have a net accumulation of nutrients, and the opposite for ecosystems with a negative NEP.

NEP is negative after forest clearing. Shading of the soil surface decreases, and soil temperature increases. Negative NEP is common in agricultural ecosystems since vegetative-carbon is continually replaced and because soil cultivation reduces the site's soil carbon stock. The soil-carbon loss varies according to the time since deforestation, the quantity and quality of the carbon stocks of the natural systems prior to conversion, and the type of agriculture practiced (Melilo, 1984).

The net nitrogen mineralization (NMIN) at the farmers' site increases after forest clearing (Melilo, 1984). This the part of the inorganic nitrogen produced from the soil organic nitrogen pool that is not used by the microbes. The warmer, wetter soil is a factor in this NMIN increase. There are also changes in soil aggregate structure (Melilo, 1984). The NMIN is available to plants or is lost due to leaching or gaseous flux. The soil nitrogen pool is eventually exhausted due to the continual removal of the easily metabolized nitrogen compounds from the site through harvest, leaching, and erosion.

In Trinidad's rainforests, about 60 kg N per ha per year is added to the forest floor annually from the 7000 kg leaf fall per ha. As much as 70% of the nitrogen in the forest floor of a Mora forest (*Mora excelsa*) can be lost when the Mora is cut and Caribbean pine (*Pinus caribbea*) planted for industrial reasons (Ahmad et al., 1982). Agriculture-derived erosion in the Caribbean has led to low total-N (ca. 0.17-0.37 % N), except for the Andepts, which have high organic-C (10-25% C) and total-N (0.68-1.50% N). Amounts of soil-N are often related positively to rainfall. Because of high rainfall fertilizing Caribbean soils has a very short lived effect on the concentration of nitrogen in the soil (Ahmad et al., 1982).

In agricultural ecosystems it is necessary to have a high enough rate of agricultural NPP so that NEP remains above zero. Nitrogen is removed from the soil in the form of harvested crop-protein. Adding manure may not be practical on steep remote slopes and promoting nitrogen fixation on the site through agroforestry may be more appropriate. The "return" of nitrogen to the land by either mechanism is accompanied by organic matter inputs and therefore leads to a positive NEP.

Agroecosystems with high structural and species diversity can exploit local nutrient resources, prevent nutrient losses, and protect crops and soil fertility. A large root system reaching deeper soil layers will promote mineral nutrient uptake. The large biomass can absorb atmospheric nutrients and reduce the mineralization rate of organic matter in the soil. There are also complementary and mutually beneficial effects with the presence of compatible species.

Natural systems have mechanisms such as nutrient immobilization in decaying plant litter (which is related to both the quantity and quality of the organic matter stocks) that allow the system to be "input insensitive." Agricultural ecosystems with reduced biomass, resource availability and homeostasis are "input sensitive" and inputs to such a system have to be related and timed to crop demands otherwise the inputs will be leached (Melilo 1984). In tropical farming a large biomass pool should be built up and rapidly turned over, so that equilibrium is maintained in the soil between buildup and breakdown of the organic matter, while ensuring a substantial increase of the CEC. Plants as standing biomass will contribute to the stabilization of microclimate and reduced erosion and organic breakdown. The organic matter content, water-holding and biological activity of the soil will increase as well (Melilo. 1984).

5. Possible solutions

a. Multi-cropping

Ancient multi-cropping systems in Central America are an example of high biomass systems that can maintain soil structure and fertility. In Tabasco Mexico, a multi-crop of maize, beans and squash had 50% higher maize yields than a maize monocrop (Gliessman, 1990). The beans in polyculture with maize nodulate more and fix more nitrogen. Net gains of nitrogen are obtained which reduces the need for fertilizer and creates more stability in the system. Beans grown with maize nodulate more and fix more nitrogen. There is less need for fertilizer and more stability in the system. Associations between pairs of species that bring mutual benefit are called mutualistic. They grow and or survive and or reproduce at a higher rate when in the presence of the other. They gain either food resources or protection from their enemies, or a favorable environment in which to grow or reproduce. Each species acts in an essentially 'selfish' manner but benefits to both outweigh the costs. The squash controls weeds since its thick, broad horizontal leaves prevent sunlight penetration. In addition leachates in rains washing the leaves contain allelopathic compounds that potentially inhibit weed growth (Gliessman, 1990). This is obviously beneficial for the farmer since it reduces labor.

b. Cover cropping

Cover cropping is also used in Central America and Mexico. The leafy plant called mucana or velvet bean forms a dense mat of interwoven vines and leaves that controls erosion, preserves soil moisture and discourages virtually all weeds. CIMMYT scientists have recorded that broadcasting mucana can help farmers reduce fallows from 5 to 2 years. The mucana also prevents the growth or shrubs and tough grasses that would have to be removed by burning. It thus saves labor during land preparation by up to two-thirds.

c. Agroforestry

Another option to reduce erosion and improve sustainability is to work directly with the small farmers in agroforestry systems that would replace shifting agriculture. Increasing the organic matter levels can help keep the CEC levels in tropical soils within their normal range 150 to 500 meq/100 g organic matter (Weischet, 1977). Standing biomass is the most practical method of storing nutrients.

Agroforestry is the intentional mixing or retention of trees in crop/animal production systems (Wiersum, 1981; Nair, 1983). It utilizes the same piece of land, either simultaneously or sequentially. The objective of most agroforestry systems is to optimize the beneficial effects of interactions of the various components, to lessen the need for outside inputs for the management of systems, and to lower the environmental impacts of farming practices. From an ecological perspective, agroforestry systems tend to share the resource pools (light conditions, water budget and soil nutrients) in a partitioned fashion. as root depths and perimeters, or location of deciduous trees in overstoreys. There are also complementary benefits when some species correct soil deficiencies for the benefit of others as in the case of nitrogen fixation by the roots of two commonly found trees *Leucaena leucocephala* and *Gliricidia sepium*. These systems also provide shadow and nutrient-rich organic matter for the underlying cultivated crops. *Erythrina poeppigiana* as an overstorey crop with coffee, as an understorey tree, is used in this way all over Trinidad and Tobago.   Simpson and Wickham (2002) found that *Gliricidia* spp. produced more leaves and thus more plant nutrients than Leucaena over a two-year period in Guyana.

Agroforestry has a historical precedent in Trinidad and Tobago and will not be rejected by farmers due to unfamiliarity. In the 18th and 19th centuries much of the Evergreen and Semi-evergreen seasonal forest in the Northern range, was replaced with cocoa by the Spanish "squatters". Later French settlers also replaced natural vegetation with cocoa. When the cocoa was planted it was common to use tannia, yams, banana and plantain as temporary shade. Immortelle trees (*Erythrina poeppigiana and E. gluaca*) are universally planted as shade trees (Beard, 1945). Under the best stands of cocoa and shade trees the soil is covered by thick leaf litter and ground vegetation is absent. When the canopy is less dense the ground vegetation is regularly cut and consists of broad leaved weeds such as *Commelina spp.* or *Chaetochloa sulcata*. These cut grasses are either left as mulch or used as fodder.

Unfortunately much of this cocoa became uneconomic or died out and vegetable crops replaced them. When these short crops are abandoned secondary bush, shrubs, bamboo and balisier (*Heliconia bihai*) take its place. The bamboo is often harvested or burnt and the balisier, although efficient in accumulating phosphorus, is insufficient to prevent erosion (Beard, 1945). In Caribbean agriculture it is still traditional to have a top storey of fruit or food-producing trees with other tiers of crops, ending with runners, such as sweet potato, or cucurbits.

d. Alley cropping

Another option for sustainable agriculture is alley cropping which provides erosion control with mulching and at the same time useful forest products and nutritious livestock feed. In a humid climate, competition for moisture between crops and trees is minimal (Kang and Reynolds, 1986). There is also a positive impact of the mulch layer formation on soil moisture conservation and the duration of this effect depends on the tree species used. Alley cropping systems have positive effects on soil organic carbon and total nitrogen contents (Kang and Reynolds, 1986). Only 25% of the foliage produced in these systems should be used for fodder. Removing more than 25% results in a lowering of the yield (i.e. from using entire prunings as mulch). Farmers with small ruminants could gain more in animal productivity than their losses in not using the entire pruning as mulch (Kang and Reynolds, 1986). Including livestock benefits soil stability because manure incorprated into the soil reduces erosion (Ekwue and Harrilal, 2010)

e. Alley Grazing

Alley grazing is simply the grazing of alley farms by livestock during periods when there are no crops. This is usually during the dry season when the animals can graze on crop residues, natural weed growth and tree regrowth. When these grazed-fallowed plots are replanted yields are 50% more than yields under continuous alley cropping, in year one. This yield diminishes gradually, becoming insignificant (12%) by the fourth year after fallow. A rotation of 4 years cropping followed by 2 years fallow provides maximum benefits. Animal grazing has to be carefully managed (Kang and Reynolds 1986).

f. Ecological constraints

Agroforestry is a more intensive form of land use than traditional forestry and a higher quantity of products may be taken out. In some instances this might cause the mineral cycle to be disturbed and in that case fertilizing may be necessary. On marginal soils the choice of suitable plant species is more crucial. It is possible that between various plant species there can be a strong competition for radiation, moisture or nutrients, or even adverse effects through chemical substances (allelopathy).

5.7. Economic factors

Agalawtte et. al. (1993) used benefit-cost analysis to evaluate sustainable agriculture (alley cropping) as an alternative to shifting cultivation in the dry zone of Sri Lanka. He found that shifting cultivation was profitable in the short run and becomes un-productive after three years. Sustainable farming is only profitable in the long run. Interventions and incentives are then needed for farmers to become conservationists. Shifting cultivators use lands where property rights are not imposed. They treat these lands as an open access resource, deriving the maximum benefits with no incentive to maintain long term productivity. Agalawtte uses figures of 4200 kg/ha/year top soil loss from monsoon rainfall on steep slopes. Conventional tillage increases the loss to 8000 kg/ha/year.

Agalawtte included intangible benefits and costs into the traditional cost benefit analysis. Difference between total revenue (output times unit price of product) and operational costs of cultivating under each system was calculated as the gross margins using current prices. Gross margins of individual crops are calculated using present output and market prices, indicating on-farm productivity losses due to soil losses and positive effects on productivity due to improvements in soil fertility and increased infiltration rates. This assumes that farm operators internalize the external cost and benefits into their decision matrix. Because shifting cultivation requires 12-15 years fallow to rejuvenate soil fertility this period was taken as the production cycle.

The net present value (NPV) of total benefits and costs over the period was calculated using a 10% discount rate. This rate is chosen because of inter-generational property rights. The scientist assumed a linear relationship between upstream soil loss and down stream sedimentation. The number of bushels of paddy lost due to down-stream sedimentation was calculated, and 10% of the eroded top soil was assumed to end up deposited down stream in the irrigation tanks.

It was also assumed that only 30% of the water in the tanks was available for irrigation. Annual soil loss was converted into volume of water displaced due to sedimented silt using a correction factor of 0.312. Each ton of soil lost in the highland was found to be associated with a loss of 20 kg of paddy (Rs.150). This was the cost of soil erosion of each of the four farming systems analyzed below.

The four farming systems analyzed were;

1. Shifting cultivation: Used for 3 or four seasons before abandonment. Productivity loss was calculated at 17%/year, making the system unproductive at the end of the 3rd year. Gross margins were calculated for a 3 year cropping cycle of two seasons per year. Soil loss is 7000 kg/ha.

2. Homegardens: 0.2 ha of multi-storey, multi species annual and tree crops. Slopes are 2% or lower, soil loss is 3500 kg/ha. A 10% productivity loss per year is used for gross margin analysis.

3. State settlement schemes in the highlands: 3 ha., clearing and contour bunded by the state. Crop rotation includes a 3 year fallow. Tree crops, seasonal and annual crops are grown. Soil loss is 5000 kg.ha.

4. Conservation system: annual crops are grown between *Gliricidia sepium* and other permanent leguminous trees. Double land area is needed to obtain the same income as shifting cultivation. Live mulching and other soil cultivation reduces soil loss to 3500 kg/ha.

Using gross margin as the net income when operational costs are deducted from the annual income, the highest per hectare GM of Rs 9265 (US $225/year/ha) comes from shifting cultivation, while conservation farming produces only Rs 4614 (US $100/year/ha). The additional operational costs for the latter system are reflected in the gross margins. Using Net Present Value (NPV) of the gross margins calculated for the 15 years shows conservation farming to be the best system. At the 10% discount rate, conservation farming has a NPV of Rs 29400 (US $750), double that of the shifting cultivation which is productive for 3 years, and unproductive for the next seven. State settlement and homegardens become unproductive by the 13th year.

The figures show that a 4 tonnes/ha. mulch rate can reduce soil loss from 7 tonnes/ha. under shifting cultivation to 200 kg./ha./year under conservation farming. The other two systems lie between these two systems. The authors estimate that labor requirements are 4 times higher in the conservation system. They estimate labor productivity under shifting cultivation at Rs 45 (US $1.0/man/day), compared to Rs 26 (US $0.5/man/day).

11. Conclusion

In the humid tropics the sustainable option may be to utilize agricultural crops that enhance soil conditions and conservation. Tropical soils need a permanent cover of trees and mulch to prevent erosion and maintain fertility and structure.

Labor intensive traditional agriculture with its emphasis on multiple cropping, multiple storeys and multiple uses are more ecologically sound in the humid tropics than conventional agriculture. However labor productivity is not high, and weed management and land preparation technologies are needed to make long term conservation as economically attractive as short term shifting cultivation. The skill may lie in mimicking existing ecosystems, elimination of slash and burn agriculture concentration nutrients on the site using various techniques. Long term education and training in sustainable techniques and land tenure are necessary steps.

References

Agalawtte, M.B., Abeygunawardena, P., 1993. Conservation farming as an alternative to shifting

cultivation in Sri Lanka: An economic alternative. Journal of Sustainable Agriculture 4, 65 -79.

Ahmad, N., Reid, E.D., Nkrumah, M., Griffith. S.M., Gabriel, L., 1982. Crop utilisation and fixation

of added ammonium in soils of the West Indies. Plant and Soil 67, 167-186.

Ahmad, N. 1991. Soil Management on Hillsides in the humid tropics, in: W. C. Moldenhauer, N. W.

Hudson, T. C. Sheng and San-wei Lee (eds.). Development of Conservation Farming

on Hillslopes by Soil and Water Conservation Society, Ankeny, Iowa, U. S. A., pp.

101-111.

Ahmad, N., Sheng, T.C., 1988. Land capability and land use of the steeplands of St. Lucia.

Organisation of American States, St. Lucia.

Anderson, M.G., Holcombe, E.A., Blake, J.R., Ghesquiere ,F., Holm-Nielsen, N., Fisseha, T.,

2011. Reducing landslide risk in communities: Evidence from the Eastern Caribbean.

Applied Geography 31, 590-599.

Anon, 2010. ODPM [Office of Disaster Preparedness and Management] Ready for Evacuation.

Newsday, Wednesday, August 11 2010.

Anonymous, 2010. Big stink in Aripero, Newsday, Friday, July 16 2010.

Asson, Cecily. 2010. Saved From Flood. Newsday, Wednesday, August 11 2010

Asson, Cecily. 2010. Snakes, caimans scare villagers. Newsday, Thursday, December 30 2010.

Beard, J.S. 1945 The Natural Vegetation of Trinidad. Oxford Forestry Memoirs No 20, Oxford

Clarendon Press.

Chenery, E.M. 1949. The Soils of Central Trinidad, Government Printing Office, Trinidad.

Ekwue, E.E., Harrilal, Al. 2010. Effect of soil type, peat, slope, compaction effort and their

interactions on infiltration, runoff and raindrop erosion of some Trinidadian soils.

Biosystems Engineering 106, 112- 118.

Ekwue, E.E., Bartholomew, J. 2011. Electrical conductivity of some soils in Trinidad as affected by density, water and peat content. Biosystems Engineering 108, 95 – 103.

FAO, 1985. Intensive multiple-use forest management in the tropics. Analysis of case studies from India, Africa, Latin America and the Caribbean. FAO, Rome.

GORTT, 2007. Chapter 12 - Natural Hazards. First Compendium of Environmental Statistics Trinidad and Tobago. CSO, Ministry of Planning and Development. POS, Republic of Trinidad and Tobago, pp. 298 – 310.

Gumbs, F.A. 1982. Soil and water management features in Trinidad and Guyana: Trop. Agric. (Trinidad) Vol. 59 No. 2. April 1982.

Gliessman, R.S. 1990. The Ecology and Management of Traditional Farming Systems in: eds, M.A. Altieri and Susanna B. Hecht, Small Farm Development. CRC Press, Baton Rouge, pp. 13 - 18.

Holcombe, E., Anderson, M., 2010. Tackling landslide risk: Helping land use policy realities in the Eastern Caribbean. Land Use Policy 27, 798 - 800.

Kang, B.T., Reynolds, L., Atta-Krah, A.N. 1990. Alley farming. Advances in Agronomy 43, 315-359.

Marshall, R.A. 1934. The physiography and vegetation of Trinidad and Tobago. Oxford Forestry Memoirs No.17. Oxford Clarendon Press.

Meersch (van der), M.K. 1992. Soil fertility aspects of alley cropping systems in relation to sustainable agriculture. Dissertationes de agricultura. Doctoraatsproefscrift Nr. 226 aan de Fakulteit der Landbouwwetenschappen van de K.U. Leuven.

Melilo, J.F. 1984. Prediction of the productivity of agricultural systems, in: F.B. Golley and J.H. Cooley (eds.), Organic Production: the relationship between agricultural and natural vegetation rates. Workshop proceedings 30 Jan-2 Feb 1984 International Assocation for

Ecology.

Nye, P.H., Greenland, D.J., 1964. Changes in the soil after clearing tropical forest. Plant and Soil XXI 1, 101-112.

Rodin, L.E., Basilevic N.J. 1968 World distribution of plant biomass, in: F.E. Eckardt (ed.). Functioning of terrestrial ecosystems at the primary production level. UNESCO. Paris. pp. 45-52.

Weischet, W., Caviedes C.N., 1993. The persisting ecological constraints of Tropical Agriculture. Longman Group UK Limited, p. 5 and pp. 259-261.

Nye, P.H., Greenland, D.J., 1960. The soil under shifting cultivation. Comm. Bur. Soils, Techn. Commun. 51. Harpenden.

Paul, Anna-Lisa (2011). Met Office: Take All Precautions. Newsday Monday, October 17 2011.

Simpson, L.A., Wickham, C.V. 2002. The performance of Leucaena and Glyricida and their potential as sources of plant nutrients in the Intermediate Savannahs of Guyana. CARDI Review No. 2: 11- 18.

Wiersum, K.F. 1994. Outline of the Agroforestry concept. Agroforestry Ecosystems Course Reader. Vakgroep Bosbouw Wageningen University.

Zimmerman, T. 1985. Conservation farming in the dry zones, in: G.R. Smucker (ed.) Planting trees with small farmers. A planning workshop, sponsored by Pan American Development Foundation and CODEPLA, Workshop Proceedings. August 5-9, 1985.

Index

Agricultural Development Bank, 10
Akaloo, 13
alligators, 19
anthroprocentrism, 2
Asa Wright, 21
Bacon, 14
Biche, 20
Block B, 14, 17
Brigand Hill, 20
buried epistemologies, 11
Bush Bush, 7, 18
Caribbean Conservation Association, 17
Caroni, 7, 15
Cascadoux, 7
cascadura, 7
Centre for Gender and Development
              Studies, 9, 32
Cross, 9, 19, 20
cultural landscapes, 21
Daniel Miller, 21
development, 9, 17
Driver, 20, 21
EIA, 14, 15
Environmental Management Authority, 10
environmental skeptics, 21
Fishing Pond, 7, 10
fortress conservation, 1
Greenpeace, 17
Guptee Lutchmedial, 14
IMA, 14
Jabar, 17
Jaikaran, 17
Jaimungal, 17
justice, 26
Karilyn Shephard, 10
Kenny Persad, 16
Kernahan, 7, 9, 15, 20
legal standing, 25
Magistrate Jurity, 17

Mahy, 19, 21
manatee, 6
Matura, 19
modernity, 21, 22
Molly Gaskin, 14, 17
moral, 2, 24, 26, 31
moral standing, 24
National Flour Mills, 10
National Wetlands Policy, 17
Nolan Bereaux, 12
Oropouche Lagoon, 7
Oropouche Swamp, 10
otters, 6
parrots, 16, 18
Pemberton, 5, 9
Plum Mitan, 7, 10, 13
Plum Mitan Women's Group, 20
Pointe à Pierre Wildfowl Trust, 14
Quesnel, 10
Ramesh Lawrence Maharaj, 7
Ramsar, 9, 14, 17
rice subsidy, 14
Rio Claro District Court, 17
San Juan Rotary Club, 14, 18
Sletto, 8
snakes, 19
Stewart Best, 12
subsistence, 10
Sukhoo, 17
Thelen and Faizool, 18
Theresa Akaloo, 10, 15
Wahab, 20
water snakes, 6
Wildlife Section, 10, 11, 12, 17, 18
wise use, 9, 21
World Bank, 12
WWF, 17